STUDIES IN HISTORY, ECONOMICS AND PUBLIC LAW

EDITED BY THE FACULTY OF POLITICAL SCIENCE
OF COLUMBIA UNIVERSITY

Number 322

CANAL IRRIGATION IN THE PUNJAB

An Economic Inquiry Relating to Certain Aspects of the Development
of Canal Irrigation by the British in the Punjab

CANAL IRRIGATION IN THE PUNJAB

AN ECONOMIC INQUIRY RELATING TO CERTAIN ASPECTS
OF THE DEVELOPMENT OF CANAL IRRIGATION
BY THE BRITISH IN THE PUNJAB

BY

PAUL W. PAUSTIAN

AMS PRESS
NEW YORK

COLUMBIA UNIVERSITY
STUDIES IN THE
SOCIAL SCIENCES

322

The Series was formerly known as
Studies in History, Economics and Public Law

Reprinted with the permission of Columbia University Press
From the edition of 1930, New York
First AMS EDITION published 1968
Manufactured in the United States of America

Library of Congress Catalogue Card Number: 68-58614

AMS PRESS, INC.
NEW YORK, N.Y. 10003

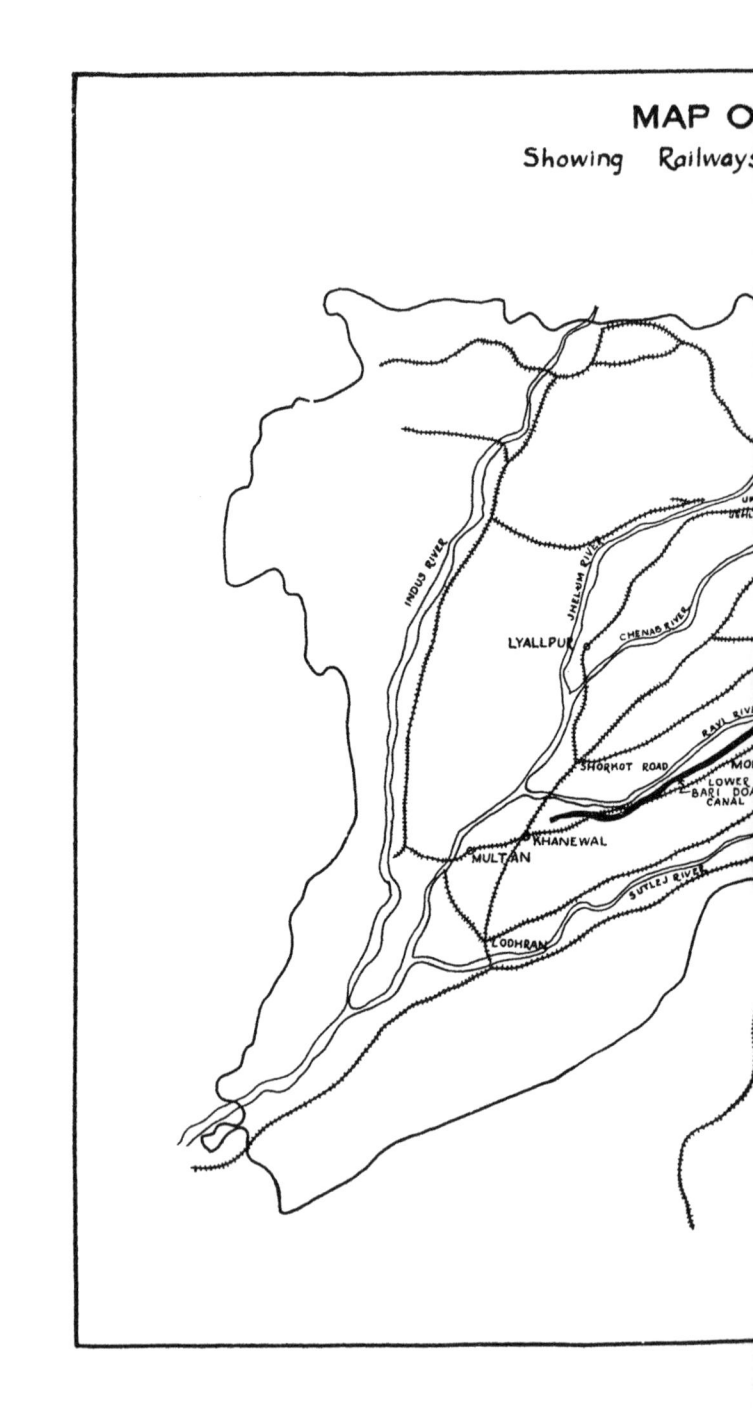

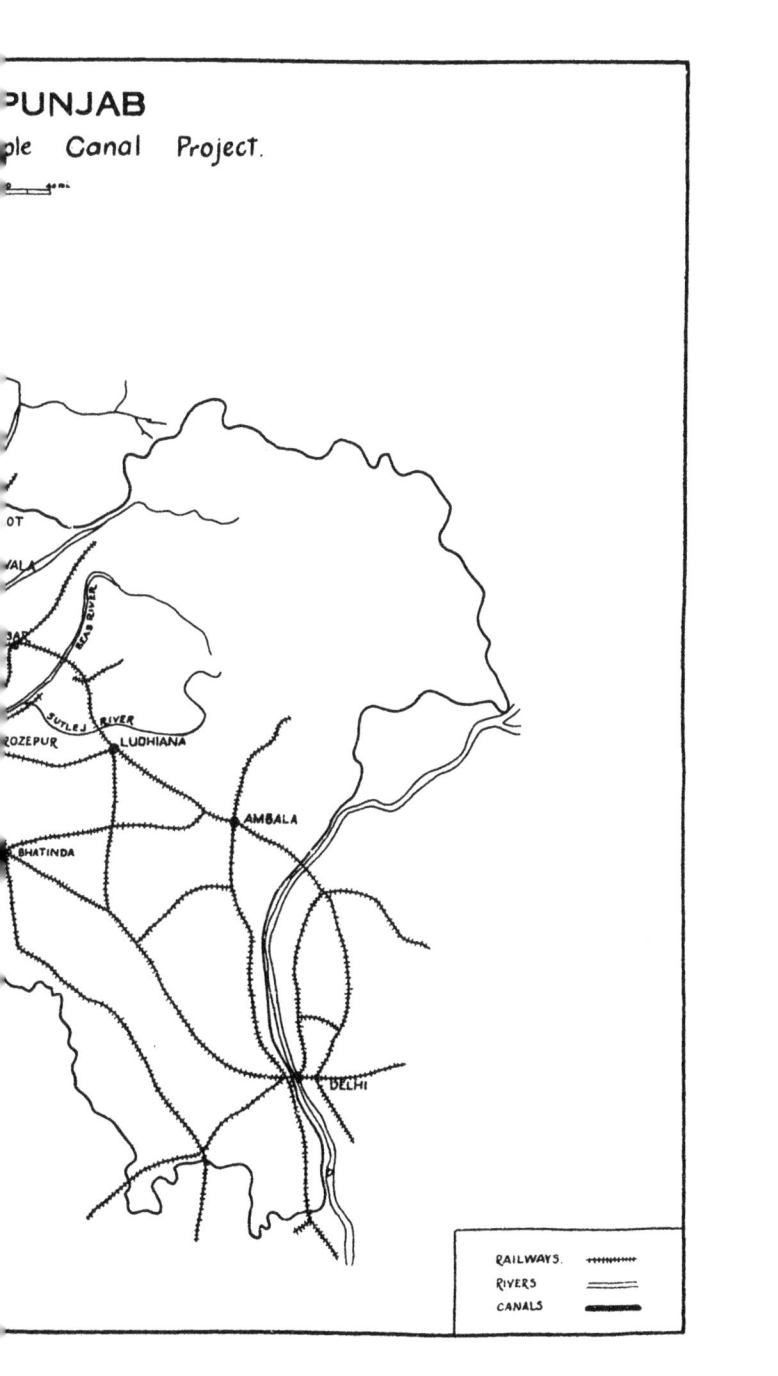

PREFACE

A study such as this perhaps requires some justification. Why attempt to discover the economic effects of irrigation in the Punjab? Are the economic problems in this portion of the world different from those obtaining in other parts where recourse to irrigation is necessary in order to reclaim additional soil for increasing populations? Why should anyone study a situation so far removed from the busy intense life of what is termed western civilization?

The incentives for attempting such a study as this may be briefly summarized as follows: Comparatively little accurate information is available to persons in countries outside of India concerning Indian economic life. For the Indian student of economic conditions materials in the form of governmental reports and special studies may exist but the reports are widely scattered and comparatively little attention has been devoted to the interpretation of the data which have thus far been collected. This monograph seeks to add to that scanty information by presenting the results of an impartial and systematic study of some of the changes which canal irrigation is bringing about in the Punjab. To even the most casual traveler through this province the intruding importance of the irrigation schemes in the economic life of the Punjab must prove impressive. The motive behind this huge irrigation development in this province has been supplied chiefly by a government whose rule has not been naturally evolved from the people governed nor amicably invited by the *Punjabis*. Hence, this study may indicate, in some measure at least, the possible result of intimate contact between western initiative and eastern economic conditions, a contact which in this case may eventually cause considerable change in the culture of the eastern people thus influenced. Perhaps this study may also indicate methods which, due to an outside influence or combination of influences, are arousing a people who seem to have been in a " quiescent, indolent, uncultivated state, with all their tastes either fully satisfied or entirely undeveloped " to put forth more of their productive energies to achieve a whole series of new desires. At any rate, such a study should provoke further inquiries to support or refute some of the chief conclusions which have been reached during the process of observation, study and recording of facts.

A few words may indicate the method of procedure. The attempt has been made to cover the literature [1] immediately relevant to the subject during a period of five years' residence in India. Considerable travel, discussion and observation in the various affected areas during four years' residence in Lahore supplemented the study of the relevant literature. Certain studies by graduate students of the Punjab University, under the direction and supervision of the writer, have afforded additional opportunities for the study of features of particular interest. It has thus been possible to check up details and points of variance discovered in the writings of others on the subject. The chief source of recorded information has been the various government reports to which frequent reference is made in this study.

No one who has attempted to study conditions amidst an environment quite foreign and about the life of a people of an alien culture will be so conceited as to place too insistent an emphasis upon the conclusions of his own study of a given situation or problem. The present monograph is submitted in the hope that it may aid in creating at least an in-

[1] *Cf.* appended bibliography.

troduction and perhaps an incentive for an exhaustive study of canal irrigation in the Punjab. Such an inquiry will probably require the cooperative efforts of a fairly large group of qualified Indian students of the subject attacking the problem by means of a number of intimate surveys and studies in the various canal colonies of the province. A splendid beginning in such cooperative research has been made by the Board of Economic Inquiry of the Punjab. A list of its publications is included in the appended bibliography.

Every effort has been put forth to state the problem as it exists. This study makes no attempt to exhaust the possibilities and implications of canal irrigation in the Punjab. This monograph is limited to a study of the historical development of irrigation during the period of British rule, 1849 to 1927, and of a few of its economic implications. The introduction sketches in broad outline the general geographical features of the Punjab, a brief characterization of its people and a reference to its early history. Part One is devoted to an historical sketch of the launching, development, construction and expansion of the canal-irrigation projects in the Province under British initiative. Part Two concerns itself with the effects of irrigation on the density and distribution of population in the province and incidentally with the relative growth of population and the extension of the arable area of the Punjab; the attempt is then made to present certain facts relating to increased food production and wealth directly due to irrigation. The trend of the argument in this section is that, due to the extension of irrigation, the available cultivable land area has increased more rapidly than the population has grown, thus resulting in a slightly broader base upon which to produce food for a relatively smaller number of people per cultivated acre. The last section deals with the fiscal aspects of canal irrigation. It is there shown that the irrigation schemes of the Government have proved a decided fiscal success and that a considerable profit has accrued. The net return to the Government from the irrigation service represents a handsome return on the capital investment. The argument concludes with a brief study of the cost of canal irrigation to the Punjab peasant. The tentative conclusion at which the study arrives is that the canal-irrigation charges which are borne by the peasant are high relative to his standard of living. It is admitted that it is at present impossible to present an accurate statement of the cost of irrigation to the peasant as definitely as has been possible in the study of the Government's return from its investments in the irrigation schemes. Up-to-date detailed scientific studies of the peasants' standard of living are not yet available. The monograph closes with the suggestion that, all things considered, perhaps the cost of irrigation to the people of the Punjab is not unduly great in view of the indirect benefits accruing to the province from the efficient management of the whole canal-irrigation project under British control.

During the period of years which have elapsed since this study was first projected materials have been collected and combined from a variety of sources. Such published records and books bearing upon this subject as have been used in specific portions of the study have been duly acknowledged in the footnotes. Certain individuals who have co-operated especially with the writer in making this monograph possible have placed him under a debt of gratitude which he desires to acknowledge: members of the staff of the Civil Secretariat Library in Lahore, India provided a most willing and pleasant aid in finding and making available for the writer the reports of the Government of the Punjab which have been so copiously utilized throughout; Professor W. E. Weld has given of his time and thought to frequent correspondence with the writer while still in India making this study; Professor V. G. Simkhovitch, by means of his constructive criticism of the plan and content of this study on the occasion of its presentation before the Graduate Seminar in Economics, greatly influenced the writer's development of the monograph along its present lines; Dr. E. M. Burns of the Department of Economics of Columbia University has carefully read the manuscript and made valuable corrections and criticisms during the period of preparation of the same; many individuals in India and in America have unconsciously aided the writer by means of conversations and interviews on the subject of inquiry. Whatever credit may accrue as the result of this study will be shared gladly with all who have made the study possible. Any blame for its shortcomings will be privately accepted. The writer is alone responsible for the point of view and general interpretation of the problems which he has presented.

P. W. P.

CONTENTS

	PAGE
AUTHOR'S PREFACE	5

PART I

HISTORICAL DEVELOPMENT OF IRRIGATION CANALS BY THE BRITISH IN THE PUNJAB

CHAPTER I
The Punjab and its people 15

CHAPTER II
Early irrigation developments in the Punjab 21

CHAPTER III
Irrigation developments, 1859–1882 33

CHAPTER IV
Rapid expansion of irrigation and the introduction of colonization schemes, 1882–1927 48

PART II

SOME ECONOMIC ASPECTS OF CANAL IRRIGATION IN THE PUNJAB

CHAPTER V
The effect of irrigation on population. 77

CHAPTER VI
The effect of irrigation on wealth and agriculture in the Punjab. . 96

CHAPTER VII

Financial aspects of canal irrigation in the Punjab 122

APPENDICES

A. Railroad mileage and date of construction 165
B. Results of land auctions in canal colonies 166
C. Agricultural experiment farms of Punjab 167
D. Canal mileage and date of construction and acreage 168
E. Areas irrigated by canals in Punjab, 1887-1927 169
F. Banks which have been located in colony towns 170
G. New canal projects and areas to be irrigated 171
H. Net return on Government investment in major irrigation canals of the Punjab, 1887-1927 172

SELECTED BIBLIOGRAPHY 173

INDEX 177

PART I

HISTORICAL DEVELOPMENT OF IRRIGATION CANALS
BY THE BRITISH IN THE PUNJAB

CHAPTER I

THE PUNJAB AND ITS PEOPLE

THE Punjab is one of the eleven major subdivisions established by the British Government for administrative purposes in India. The bulk of this province lies between the thirtieth and the thirty-third degrees north latitude. Its northern boundary is a series of mountain ranges belonging to the Himalayan system; these ranges also form the northeastern limit of the province which also impinges upon the United Provinces on the east; to the south and southwest the Rajputana and the Sind deserts respectively form an unfruitful boundary to the Punjab while on the northwest portion the province is bounded by the Frontier Provinces. Thus the Punjab is quite severely landlocked by mountains and deserts except for that small portion of its boundary which stretches out toward the United Provinces. This quite natural isolation of the province has permitted the development of a fairly unified culture in this section of India and is largely responsible for much of the social solidarity of the people of the Punjab.

The total area of the Punjab and its dependencies is 133,741 square miles, of which, 97,209 square miles are British territory, while native states of various areas account for the remaining 36,950 square miles. It is interesting to note that these Indian States within the Punjab are completely surrounded by British territory within the province. The province gets its name from five rivers, the Indus, Jhelum, Chenab, Ravi and Sutlej, which course through its comparatively level plain. From time immemorial this section of

India has been called the " Land of the Five Rivers."¹ This vast alluvial plain, which Mr. M. L. Darling calls " the real Punjab "² is the most striking feature of the geographical makeup of the province. Approximately three-fourths of the total area of the Punjab lies within this gently sloping plain. A glance at a map of the Punjab indicates that these rivers flow more or less parallel from northeast to southwest finally uniting their sluggish waters to increase the volume of the Indus River near Kashmor in the southwestern corner of the province. The mountain ranges on the northern and eastern boundaries of the area form a picturesque and well nigh impregnable natural line of defense while acting also as an important watershed against which the moisture-laden winds are forced to deposit, in the form of rain or snow, the water which later courses through the " land of the five rivers ". The submontane tract separating the plain proper from the mountainous portion of the province breaks the otherwise monotonous terrain and presents soil and climatic conditions quite different from those obtaining on the plain.

The winters are comparatively cold and the summers are long and hot. Its inland position in connection with its sandy soil and its proximity to the Rajputana and Sind deserts gives the Punjab a climate subject to great extremes of temperature. Thus, the winter of a temperate climate is followed by a truly tropical hot season. Such a climate tends to breed a hardy martial race as various writers are

¹ The word Punjab, sometimes spelled Panjab, is derived from two Persian words, *panj* meaning five, and *ab* meaning water. Hence the Punjab is the land of the five waters.

² Darling, M. L., *The Punjab Peasant In Prosperity And In Debt* (London, 1928), p. 84. This Punjab plain is practically level. Its altitude is slightly over 1000 feet at the base of the foothills in the north and northeast, and about 700 feet in the southwestern corner of the province. The slope of the land on which most of the canals have been constructed is between two and three feet per mile.

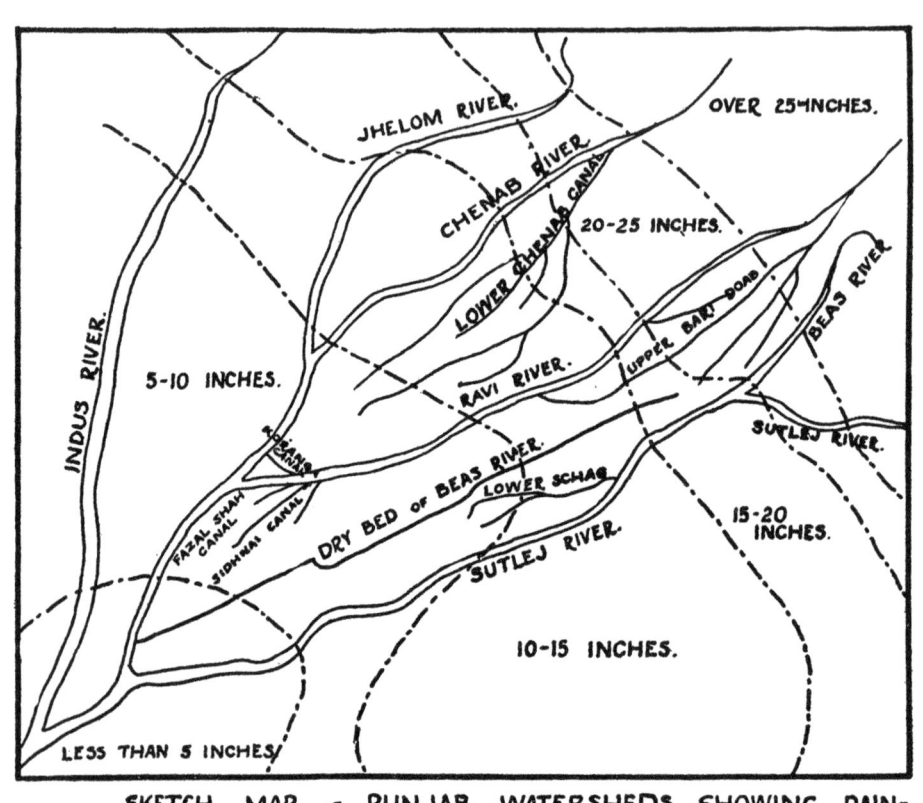

SKETCH MAP of PUNJAB WATERSHEDS, SHOWING RAINFALL CONTOURS. (Facing p.132 - Land of Five Rivers. vol.I. 1921-1922.).

fond of recalling.[1] The *Punjabi*, is a man of splendid physique, long inured to the extremes of climate which he must survive or be eliminated. The World War found the Punjab peasant turned soldier under the banner of his British conqueror quite capable of enduring a winter in the mud of Flanders or a summer amidst the sands of Mesopotamia. The *Punjabi* physique is not equalled by that of any of the other peoples of India.

Agriculture is the predominant activity of its people. Outside of the railway shops in Lahore and a few scattered flour and cotton-ginning mills located in various sections of the province, agriculture is the one industry in the Punjab. Agriculture, however, depends upon moisture and the rainfall is too precarious to permit of a thriving agricultural population without some artificial means of stabilizing the water supply. In the submontane areas the annual rainfall amounts to from twenty-five to thirty-three inches, but the moisture shades off to a precarious five inches or less in the vicinity of the conjunction of the Sutlej and Indus Rivers. Lahore receives an annual rainfall of approximately fifteen to twenty inches. The region surrounding Multan averages less than five inches.[2] Evidently agriculture is a highly speculative venture under such conditions. In the greater part of the Punjab where irrigation is required and possible the annual rainfall is about twelve inches and most of the rain falls in July, August and the first half of September. This militates against a prosperous agriculture since the rain comes too irregularly and during too brief a portion of the year to permit advantageous utilization of the otherwise favorable climate during the remainder of the year.

The Punjab, lying in the direct path of any invaders of India coming through the passes on the northwestern

[1] Trevaskis, H. K., *The Land Of The Five Rivers* (London, 1928), p. 1.

[2] *Cf.* Map of Punjab watersheds showing rainfall contours facing this page.

frontier, has had a varied and exciting history. Due to the relatively late development of efficient ocean transportation facilities on a large scale India was quite amply protected by natural isolating influences from invasion by sea on two sides of its vast triangular expanse, while on the north the Himalayas proved equally solid barriers against possible invaders of Hindustan. Only in the northwest were there available mountain passes lying comparatively open to incursions of warlike peoples of western and central Asia. The Khyber and Bolan passes were the two open doors to the vast plains of Punjab which thus proved to be the highway over which the invaders came to plunder and despoil India. Hence, it is not surprising to find the people of the Punjab relatively more warlike, of better physique, less restrained by custom and tradition, slightly more pliable as to their social and political organization and of somewhat mixed racial stocks. For each succeeding wave of invaders caused some change or adaptation in the cultural environment of the people of the Punjab.

Were it not outside the province of this inquiry, a careful study of the effects of the Aryan and Scythian immigrations would prove both entertaining and instructive. Similarly, the chronicling of events relating to Alexander's historic invasion of northwest India, of the relics and reminders of Greek influence on some of the early Indian cities like Taxila and Harappa which are now being excavated by the Indian Archeological Department; the saga of the Mohammedan invaders who founded the greatest Islamic Empire of historical record in India; the splendidly heroic and thrilling adventures of the successive invasions led by Mahmud, Tamerlane, Nadir Shah and Ahmad Shah; the events and amazingly intriguing cunning of Baber and of Humayun, though interesting lie beyond the scope of this study. These chapters in the early history of India and the Punjab must be omitted from this monograph. The people of the Punjab would thus appear to be the result of a long period of complex evolution in which a large number of strangely intermingled elements of many races and cultures have been enmeshed. Our interest centres however upon some of the contemporary economic problems growing out of an even more recent invasion.

In the latter half of the eighteenth century, because of the decline of the Mohammedan power, smaller political units were established in various portions of the tottering Empire of the Moguls. The Maratha and Sikh groups attempted to reassert the ancient Aryan hold upon India. During the period just preceding 1759 the Marathas under Raghoba occupied most of the Punjab.[1] By 1799, however, the Sikhs under Maharajah Ranjit Singh had acquired possession of the city of Lahore and were rapidly extending their kingdom.[2] Ranjit Singh was ambitious to annex by successful war territories lying south of the Sutlej River but was stopped by Lord Minto in 1809 by the establishment of "perpetual amity" between the Sikh Kingdom and the British Governor-General. But Ranjit Singh continued to consolidate and strengthen his hold upon the government of the Punjab and extended his territory steadily toward the northwest. Upon the death of the Maharajah Ranjit Singh in 1839, the Sikh Kingdom included Kashmir and Kangra within its boundaries. A period of violent anarchy followed during which British territory was rashly invaded with the object of plundering Delhi in 1845.[3] The brief war which followed resulted in a British victory. The terms of peace included joint occupation of Lahore by the British and the Sikhs under the supervision of Sir Henry Lawrence who was appointed British Resident. This make-shift arrangement lasted until 1849 when, after the murder of two British officials at Multan by the Sikhs and further outbreaks of violence, the Punjab was formally

[1] Smith, V., *Oxford History Of India* (London, 1923), 2nd edition, p. 460.
[2] *Ibid.*, p. 614. [3] *Ibid.*, pp. 690-700.

annexed to British territory. Since that time the province has been under British government.

Prior to the British conquest of the Punjab, no government not in possession of the Afghan hill country had been able to hold the plains for more than a brief period. Command of the Afghan hills and the passes into India was strategically desirable for the control and government of a stable Punjab state. Hence the Punjab of history has been more influenced by the inhabitants of those plains of Central Asia than by its Indian neighbors, who except for a brief period under the great Maratha leader, Sindhia, have never had the power or even the wish to attack the hardier and more manly races of the Punjab.[1] The nomadic life of the people centred about their flocks and the moving camps. The soil was free and the grasses common property at least until it was grazed. Hence a quasimilitary patriarchal system was gradually developed to control the group life of the Nomads.[2] Kinship was the only political bond known to such a society. Roving from grazing ground to grazing ground, living a free life in the open, such societies achieved a splendid adaptation to the nomadic life of Central Asia. They were mobile, their members tended to increase, the strenuous outdoor life eliminated the unfit, and thus these roving groups presented in cases of necessity a dangerously effective fighting unit. When drought made pastures scarce and it became necessary for the whole aggregation to storm the passes into more plentiful pasture lands to the southeast fighting and conquest became the law of survival.[3]

In such terms various writers on the early history of the Punjab attempt to account for the invasions of the Punjab in the past.

[1] Trevaskis, H. K., *op. cit.*, p. 12.

[2] Maine, Sir H. S., *Village Communities In The East And West* (London, 1876, 3rd edition), p. 54. *Cf.* also *Cambridge History of India* (London, 1922), vol. i, p. 38 *et seq.*

[3] Trevaskis, *op. cit.*, p. 16 *et seq.*

CHAPTER II

Early Irrigation Developments in the Punjab

THE British conquerors of the Punjab lost no time in carefully surveying the economic possibilities of the newly acquired territory in the " land of the five rivers ". One of the first features noted was the presence in the territory of various irrigation schemes dating from fairly ancient times. The new Government's interest in the possibilities of utilization and improvement of the existing irrigation works and the projection of new schemes is evidenced by the following quotation from the first report on the general administration of the Punjab:

The capabilities of the Punjab for canal irrigation are notorious. It is intersected by great rivers; it is bounded on two sides by hills, whence pour down countless rivulets; the general surface of the land slopes southward with a considerable gradient. These facts at once proclaim it to be a country eminently adapted for canals. Nearly all the dynasties which have ruled over the five rivers have done something for irrigation; nearly every district possesses flowing canals or else ruins of ancient water courses; many of the valleys and plains at the base of the Himalaya ranges are moistened by water cuts conducted from the mountain torrents. The people, deeply sensible of the value of these works, mutually combine with an unusual degree of harmony and public spirit, not only for the construction of the reservoirs, but also for the distribution of the water and the regulation of the supply. In such cases, when the community displays so much aptitude for self-government, the Board consider non-interference the best policy, while they would always be ready to afford any aid which might be solicited.[1]

[1] *Punjab Government General Administration Report 1849-1850* (Calcutta, 1850), p. 133.

The chief irrigation works which the British found operating upon annexation of the Punjab were the inundation canals along the Jumna River, inundation works along the Indus and the lower Chenab Rivers, and the Huslee Canal which was so constructed as to provide a perennial flow of waters from the River Ravi to the city of Lahore and to the holy city of the Sikhs, Amritsar. These works will be described shortly.

The fact that the people of the Punjab realized the importance and the advantage of irrigation schemes and were capable of successfully developing and administering, on their own initiative, certain irrigation projects on a small scale encouraged the British officials in the Punjab to propose certain immediate developments in the way of new irrigation schemes. They found the soil thirsty and mountain streams carrying the water of the melting snows during the hot season wastefully to the sea. Hence among the very first matters which the civil engineers had to consider was the taking of levels with a view to the improvement and better arrangement of irrigation. The *Punjabi* peasant whom the new conquerors found in the annexed territory was apparently an industrious and capable individual who devoted his time seriously to the business of winning a living from the soil under none too favorable conditions. The peasants were at that time engaged not only in producing the minimum necessities for existence but were exporting agricultural products such as sugar-cane and indigo to Sind and Kabul. Cotton was also produced but the uncertainty of the seasons prevented the development of cotton culture on a profitable basis prior to the advent of the British. Wheat and maize were the two staple crops grown and they were declared to be of excellent quality.[1] The general methods employed in agriculture were fairly good, manures were used and the ad-

---

[1] *Administration Report, op. cit.*, p. 7.

vantage of rotation of crops was recognized. Canals were not infrequent and well irrigation, by means of the Persian wheel, was widely practised in the territory,[1] but the chief obstacle to a profitable agriculture was declared to be the lack of sufficient and regular rainfall.[2]

British experience in other portions of India had proved to the officials in charge of the various sections governed that India was a land peculiarly susceptible to famine. The Punjab was similarly liable to periodic famine, since its people eked out an existence precarious at best from one pursuit, agriculture, which was dependent upon the vagaries of the weather for the production of the necessary food supply for the people. Famine in the Punjab was inevitable as long as the bulk of the people were thus dependent upon agriculture which in turn depended upon a very uncertain rainfall. The unirrigated soil of the Punjab except in the submontane districts received insufficient rainfall badly distributed over the growing season and hence produced scant food supplies. And whenever the scanty rainfall failed, famine resulted, since the produce of the good years failed to provide a margin of surplus sufficient to tide the population of the province over the period of drought. When the rain failed, not only was the production of food curtailed but the peasants were deprived of even the precarious credit which the money lenders extended to them in ordinary times. Hence, without credit, food or exchangeable commodities with which to barter, and with transportation facilities and means of communication in general appallingly limited, the Punjab peasant during periods of famine was unable to secure from outside the province the means of subsistence to tide him over the famine period. The Punjab was visited by severe famines in 1783, 1802, 1812, 1817, 1824, 1833, 1837. In 1841 an epidemic of fever reached such serious proportions that the crops died

[1] *Administration Report, op. cit.*, p. 7.   [2] *Ibid.*, p. 8.

standing for lack of harvesters, precipitating a famine, which however, was not due to lack of rainfall. Erratic monsoons again caused serious famines in 1851-2, 1860, 1868-69 and 1877-78.[1] Since 1878 famines in the Punjab have not reached serious proportions because of the growing efficiency of the irrigation-canal system. In 1901 the Indian Irrigation Commission made a minute study of the rainfall experience of the Punjab and on the basis of that study indicated that in the sub-montane area of the province with an average rainfall of thirty-three inches, out of every fifty years there would probably be ten dry years of which four would be years of severe drought. The Southeast Punjab with an average rainfall of approximately ten inches might expect out of every fifty years thirteen dry years of which five would be years of extreme drought. The West Junjab with an average rainfall of approximately ten inches might expect in every fifty-year period,[2] fourteen dry years of which nine would be years of severe drought. One of the most severe famines on record in the Punjab occurred during the period of the Sikh ascendancy in 1783.

In the east of the Punjab the country was depopulated, the peasants abandoning their villages and dying in thousands of disease and want; the country swarmed with thieves and highway robbers, and the state of anarchy was almost inconceivable. In the fertile and populous central districts the seeds of the acacia and cotton plant were greedily devoured; so many died of starvation that bodies were thrown into wells unburied, mothers cast their children into the rivers, and even cannibalism was resorted to. The cattle nearly all died, or were eaten up by the

---

[1] For a study of Indian Famines, see, *Punjab Government Famine Reports*; Baine, J. A., *Famines in India*; Blair, C., *Indian Famines*; Loveday, A., *History And Economic Of Indian Famines*; Roy, S., *An Essay On The Economic Causes Of Famines*.

[2] *Government of India, Indian Irrigation Commission Report*, 1901-3, (Calcutta, 1903, vol. i, pt. 1, p. 4.

starving Muhammadans. Many ruins of villages are traceable to this famine, and extraordinary friendships grew up among the survivors, who clung together sharing everything available as food. The famine was followed by great mortality from fever and ague, and a large proportion of those who had escaped starvation fell victims to disease.[1]

These recurring famines were due chiefly to three causes: a population which pressed too heavily upon the available land for subsistence; a lack of rainfall which precluded the possibility of supplying the minimum of food required for subsistence during the periods of drought; lack of transportation facilities and means of communication. Famines thus tended to create a balance between population and food supply. The British, however, sought to avoid the recurrence of famines by launching upon an extensive development of canal irrigation to remove the extreme reliance upon the shifting rainfall. Thus during the first year of British occupation of the Punjab the Baree Doab Canal was projected and construction begun. The lack in transportation facilities was also to be remedied by the proposed developments, since the irrigation canals were to provide water transportation facilities and along the banks of the canals there were to be constructed highways. Railway construction was also projected shortly, especially following the experiences of the British during the Indian Mutiny of 1857. Railway construction was begun in the province in 1858; the first train was operated between Lahore and Amritsar in 1861. The development of transportation facilities, railways, canals and highways shares with the steady expansion of irrigation canals the honor of changing the Punjab from a constantly famine-endangered province to one which is practically secure from such visitations.[2]

[1] Ibbetson, D. C. J., *Punjab Census Report*, 1881, vol. i, p. 117.
[2] A table indicating the extent of railway development in the Punjab will be found in Appendix A. *Cf.* frontispiece map showing railways of the Punjab.

The one perennial canal which was operating when the province was annexed by the British in 1849 was the old Huslee Canal which had been built by Nawab Ali Khan at the command of the Emperor Shah Jehan in 1633. This canal took the waters of the River Ravi at Madhopur and carried them to the palace gardens at Lahore. These gardens have been quite completely restored by the Government of the Punjab and furnish the show-places par excellence of the capital of the Punjab today. During the Sikh regime the Huslee Canal was extended from Lahore to Amritsar to fill the tanks surrounding the Golden Temple in that famous holy city of the Sikhs. At the time of annexation this canal was producing a net income for the Government of Rs. 76,000 collected in the form of water rates from lands irrigated at the rate of Rs. 2-6-8 per acre.[1] The purpose of the Huslee Canal was to convey a moderate volume of water to Lahore at a minimum expense and any irrigation which was made possible en route was of purely secondary importance. It was generally so constructed as to avoid the table land, following the natural line of drainage for a tortuous course of 110 miles. It varied in width from five to fifteen feet and in depth from two to seven feet. The British engineers who first considered the further development of irrigation in this section of the province soon decided that this canal could not be extended profitably to furnish extensive irrigation since it could not be made to reach the level of the higher-lying lands which most needed moisture. Hence the ultimate abandonment of this old irrigation work was decided upon, although during the period of alteration of the whole irrigation scheme in this sector, repairs were made which made it possible for the area irrigated to be slightly increased. Eventually this area was to be watered from an

[1] *Administration Report, op. cit.*, p. 9. Rs. 2-6-8 should be read Rupees two, annas six, pies eight.

entirely new irrigation canal which was planned during the first year of British occupation of the Punjab.[1]

Work was thus immediately begun on the new surveys and the projection of irrigation schemes which were to change the economic significance of the entire Punjab. Not only were the British quick to see the advantages which might accrue to the Government from the development of irrigation schemes to improve the economic condition of the Punjab peasants, but there was a poignant need of finding some kind of work for the disbanded Sikh soldiers who, immediately after annexation were demobilized and left without any settled mode of earning a living in a peaceful manner. And incidentally, this demobilized force of Sikh soldiers provided a ready supply of fairly efficient workmen to help in the excavation of the new projects. It was quite natural that the first new scheme should center in the portion of the territory known as the *Baree Doab,* the land lying between the River Ravi and the River Beas. This was the territory through which the Huslee Canal had been built previously. The officials thus wrote in their first report:

No part of the new territory is so important politically and socially as the Baree Doab. In no *doab* is there so much high land susceptible to culture; so many hands to work; so fine a population to be supported. In the upper or Manja division, smaller canals had been constructed and worked by successive Governors for several generations. In the lower division, the central waste is interspersed, not only with ruined cities, but also with relics of canals and aqueducts. . . . The Resident at Lahore studied the feasibility of enlarging the Shah Nuhur or Huslee Canal which intersected the upper portion of the *Doab.* By permission of the Right Hon'ble Lord Hardinge, the then Governor General, Lieutenant Anderson, Captain Longden and Lieutenant Hodson were deputed, under Lieutenant Colonel

[1] *Administration Report, op. cit.,* pp. 135-136.

Napier's own supervision to survey and examine the line. These local inquiries were interrupted by the Mooltan insurrection but not until a considerable portion of both the Upper Baree and the Rechna Doabs had been roughly, but scientifically, examined, and Colonel Napier had devised schemes for two great canals, one from the Ravee and the other from the Chenab River.[1]

It was decided that the old Huslee Canal should be superseded and a larger new canal with three branches conducted through the entire length of the *Doab*. The central line of the new canal was to be 247 miles long. It was to commence at the point where the Ravi River debouches from the lowest of the Himalayan ranges and thence cutting through a high bank it was to be carried across two mountain torrents until it should have gained the table lands. Then following the general watershed in the *Doab* it was to pass near the towns of Deenanugur, Butala and Amritsar, pass through what was then barren waste land, studded here and there with mounds signifying buried cities of an ancient day, revivify the desert and rejoin the Ravi River about 56 miles above Mootlan. One of its branches was to take off at the thirtieth mile of the main canal and reach out toward the ancient city of Kusoor and from this branch a smaller branch was to carry water to the eastward toward the Sutlej River. Still further down, the main channel was to give off another branch which was to spread fertility toward and about the city of Lahore. Thus, this projected canal was to aggregate some 466 miles of canal channel. Not only was this canal to provide irrigation but for most of the main canal's length it was to provide navigation for the flat-bottomed native boats, thus providing an easier outlet for the surplus produce which was to be raised in the areas irrigated.[2]

[1] *Administration Report, op. cit.*, p. 134.
[2] *Ibid., op. cit.*, p. 136 *et seq.* See map facing p. 48.

EARLY IRRIGATION DEVELOPMENTS

Except for the upper reaches of the main canal channel, the construction problems confronting the engineers were comparatively simple since the soil in the *Doab* was not rocky; it sloped quite gradually toward the south, the gradient being as little as four feet per mile for some fifty miles below the first rather steep descent from the foothills. Thus the construction problems were not difficult. Indeed, the Punjab offered an almost ideal base for the development of irrigation schemes. The Himalayas provided a natural reservoir of water which flowed down the rivers during the very portion of the year when moisture was most needed on the parched plains. Hence, the construction of headworks in the form of elaborate and expensive masonry dams was not necessary. Furthermore, the few masonry works that needed to be built were constructed from rock and stone immediately available in the hill country. The very gradual gradient of this portion of the Punjab made possible the entertainment of high hopes for the navigation schemes which were to be developed as a by-product of the irrigation projects proper. It was felt that flat-bottomed boats constructed so as to navigate in two feet of water, could successfully ply up and down the canal and thence into the River Ravi and eventually transport the excess produce of the province to the sea for export. Prior to the construction of this canal the plan of canal alignment pursued by the pre-British canal builders directed the channels of the canals along the natural water line or depressions. The new canal was, as stated above, aligned along the ridge of the *Doab,* thus greatly increasing the area which could be successfully irrigated by natural flow. The natural contour of the topography of the Punjab thus made possible a relatively easy and very efficient canal irrigation development.

Another new venture included in the original plans for the canal through the Baree Doab was the plan to plant trees in the form of groves along the banks of the canal. These were to provide shade trees, which had quite disappeared from the countryside, as well as a source of fuel for the people of the province. Accordingly a space of 300 to 400 feet along the banks of the canal and its branches was set apart for avenues of trees. These trees were to provide natural strength to the canal banks which in certain places rise to considerable heights above the land to be irrigated.[1] A still further source of profit from the new canals was visualized by the engineers and officials sponsoring the new project in a scheme to sell or lease water rights to mills which were to be built at certain points in the canals where it was necessary to build masonry dams or rapids to control the steady flow of water. Hence, the canal was expected to yield revenue from the collection of water rates; rents from mill sites which used water power; charges for transportation using the canal as a toll highway for native boats carrying agricultural and other produce; and from the sale of waste lands which with irrigation facilities would become valuable agricultural land productive of increased food supplies for the population. It was also realized that in addition to these direct benefits from irrigation which would accrue to the Government in the form of these specific forms of revenue, the development of agriculture in these areas would tend to increase the growth of cities and towns and thus would perhaps prove to be the base on which more people could be gainfully employed thereby presenting to the Government a larger source of taxable income.[2]

The entire outlay at an average cost per mile of Rs. 21,546 on 247 miles was estimated at 53 lakhs [3] (or £ 530,000 sterling); while the annual net revenue was estimated at 14½

---

[1] *Administration Report, op. cit.,* pp. 138 *et seq.*

[2] *Ibid.,* p. 138.

EARLY IRRIGATION DEVELOPMENTS

lakhs[1] or equivalent to 27½ per cent on the capital investment. Thus the engineers suggested that the canal would be able to repay its cost of construction in about five years after it was fully completed and in efficient operation. The details of the estimated net return per year to Government may be thus indicated: From the experience gained in other irrigation areas in India, it was estimated that each cubic foot of water per second flowing through the canal would irrigate 218 acres in the course of a year. Hence the 3000 cubic feet of water, the estimated volume of the projected canal, would irrigate 654,000 acres. At the established rate charged on the Huslee Canal areas at the time of British annexation of the Punjab, of Rs. 2-6-8 per acre, the yield per year would be Rs. 15,80,500.[2] Mill rents were estimated at another Rs. 50,000 per year, freight duties collected on the canals were expected to yield another Rs. 20,000, while Rs. 11,000 was estimated as the probable income from other canal produce, such as trees, timber and fuel, making a total estimated gross income of Rs. 16,61,500 per year. Deducting the sum of Rs. 2,00,000 for estimated current expenses of operation, the net revenue was estimated at approximately 14½ *lakhs* of rupees per year. Furthermore, the lands lying just beyond the irrigated areas would also receive an indirect benefit of less easily measurable dimensions which would eventually be reflected in an enhanced land revenue collected by the Government. On the basis of such an optimistic estimate on the part of the engineering department the British launched the new irrigation project before the close of the first year of occupation of the province. By the end of the year 1850, the second year of occupation, work on the

---

[1] A *lakh* is equivalent to 100,000. *Ibid.*, p. 138. In 1850 one rupee was equivalent to one-tenth of a pound sterling.

[2] *Administration Report, op. cit.*, p. 139. The figures Rs. 15,80,500 should be read Rupees, fifteen lakhs, eighty thousand, five hundred.

canal had been begun throughout the first thirty miles of the main channel; some of the engineering difficulties connected with the crossing of certain mountain torrents by the main channel had been successfully solved. Within five years it was hoped the entire project might be completed. Just how the canal was to change the area through which it was being built is indicated in the following:

This canal will preserve, from uncertainty of season, and from the chances of periodical drought and even famine, a tract whose inhabitants are the very flower of the nation. It will also restore animation and fertility to a tract which was once the abode of men, and the scene of commerce and agriculture, but which, through the evolution of the centuries has become a haunt of wild beasts, a wilderness of woods and brushwood, rendered even more desolate by the appearance of ruins and relics, the sad tokens of banished prosperity. . . . The Board believes that it will be the pride as well as the interest of the British Government to originate and carry out such a work as this.[1]

The argument of this chapter shows that immediately upon gaining control of the Punjab the British Government took steps to change the economic condition of the province. Its interest in the possibility of change was probably increased by the encouraging estimates of probable future revenue-producing power which the canal irrigation schemes could be expected to yield in the future. But humanitarian motives were also present as is indicated by some of the references to improving the peasants' condition in the report of various Government officials during the year 1849-50, the first year of British control and ascendancy in the Punjab. In the chapters immediately following the further development of irrigation in the Punjab will be sketched.

[1] *Administration Report, op. cit.*, p. 140.

CHAPTER III

Irrigation Developments, 1859-1882

THE New Baree Doab Canal proved to be a bigger construction project than the engineers' preliminary estimates indicated. Construction work was begun, as shown in the previous chapter, shortly after the annexation of the Punjab in 1849. On April 11, 1859, water was first admitted into the main channel. The Mutiny caused some delay in the prosecution of the work in 1857-58 but during the latter portion of 1858 work was resumed on its former scale. The original estimates of the cost of the project had by this time been increased from Rs. 53 lakhs to Rs. 135 lakhs due to engineering difficulties caused chiefly by floods in the sections of the work near the canal head. By 1860 progress had been made on 287 miles of the Baree Doab Canal and 114 miles of the main channel were in operation. Less than one-tenth of the excavation remained to be completed. However, the 114 miles of channel through which water was then being carried were not ready for navigation because of the postponement of construction work on several locks. The hope was expressed by officials at the time that these might soon be completed so that invalid English soldiers might be able to proceed from Lahore to the hills in comfort.[1] Naturally, the project's revenue-producing powers were also lagging behind the original estimates on account of the longer time required for the completion of the work.

On the other hand, the old Huslee Canal continued to be

[1] *General Administration Report*, 1859-60 (Calcutta, 1860), p. 85.

worked at a profit to an extent that astonished the officials in charge. The income from the Huslee Canal in 1859-60 amounted to Rs. 94,250 out of which maintenance charges came to Rs. 29,039, leaving a net revenue of Rs. 65,201 for the Government. Other of the older works which the British were renovating during this period were the inundation canals along the Indus and the Sutlej Rivers. The inundation, or water-level canals were dependent for their flow upon the level of the water in the rivers, and were generally useful only during the flood periods. These canals were not trustworthy sources of an assured water supply because of the variation in the floods in successive seasons. Furthermore, the areas irrigated by the inundation canals were limited to the lowlands in the river valleys. They provided no means of lifting the water from the rivers to the higher land contiguous to the rivers, except where crude lever pumps, similar to those used in the Nile valley, were utilized. The inundation canals were simple projects which required a minimum of construction work. In contrast to these crude inundation canals, the perennial canals, of which the Baree Doab, then under construction, was to be the first,[1] were planned and executed by British engineers according to a scheme which would permit canal irrigation practically during the whole year and especially during the long hot season when water was absolutely necessary to the growing crops.

As indicated in the previous chapter, another system of canals which the British upon gaining control of the Punjab found operating on an inundation basis was the series of canals along the Jumna River known as the Western Jumna canals. These canals along the Jumna earned a clear profit of Rs. 2,00,000 in 1859-60, representing for that period a

---

[1] As shown in the previous chapter, the Huslee Canal was the only perennial canal which the British found operating in the Punjab upon annexation of the territory.

remarkable government revenue. The oldest of the works found in the Punjab was the Western Jumna Canal built by Feroze Shah to convey water to the lodge of the Emperor at Hissar. The construction of this canal dates from the 14th century. Akbar, in 1568 renovated this canal but it had fallen into disuse in the early part of the 18th century except for small portions which continued to be locally worked. The old canal, according to the custom of former days, was constructed along lines of natural drainage levels instead of on the ridges of the country. This led to the formation of swamps and water-logging which bade fair to overbalance the benefits of irrigation. It is true however that with a minimum of expenditure, the British were able to use many of these old canals as a part of the renovated Western Jumna system.[1] The presence of this water-logging hazard in the regions served by the Western Jumna canals caused serious attention to be devoted to the problem as early as 1859 as the following indicates:

Fiscally, nothing could be more satisfactory: unhappily there is another side to the picture. Unscientifically constructed, the canal bed is in many places above the level of the country, and interferes with its natural drainage. Swamps are formed, the soil is deteriorated: worst of all, with the excess of water, up comes from below a coating of salt which has for several years past gone on spreading, and has unquestionably not only injured the productive powers of the land, but impaired the physical condition of the people. The Lieutenant Governor has reduced the land revenue, wherever deterioration of the soil has been proved, and doubtless much relief has been afforded; but a permanent remedy will probably be found only in an extensive system of drainage, which Captain Turnbull is now devising.[2]

[1] Thompson, W. P., *Punjab Irrigation* (Lahore, 1925), pp. 1 *et seq.*
[2] *Administration Report, op. cit.*, paragraph 94.

The successful remedy for this problem of water-logging which was thus to have been devised by the engineers in 1860 has not yet materialized. The evil is becoming increasingly distressing as is indicated by the serious attention given to this problem by the members of the Royal Commission on Agriculture during its study of agricultural problems in the Punjab in 1926.[1] Approximately one per cent of the area irrigated in the province in 1926 was threatened by the evil of water-logging. This is one of the most serious problems confronting the Irrigation Department in the Punjab. It has received increasing attention on the part of Government since its first inclusion in the report on the adminstration of irrigation in the Punjab in 1860.

As indicated earlier in this chapter, the Baree Doab Canal was opened for irrigation in the sections completed in 1859. In that first year of this canal's operation, 50,505 acres were irrigated, and the Government's net revenue from this source was Rs. 2,26,876. From that year onward, the Baree Doab Canal proved an increasingly important source of revenue for the Government as will be shown in Chapter VII. The inundation canals along the Indus during the fiscal year 1860-61 irrigated 65,889 acres while the Government was aiding the *zamindars* in clearing some of the old canal channels and generally increasing the efficiency of this portion of the indigenous system of irrigation. Thus the acreage devoted to crops under irrigation was expanding and the officials of the day were becoming alarmed at the fall in prices of agricultural products which was occurring. The increase in the acreage available for the production of food crops tended to increase the amount of such crops produced resulting in a fall in prices for agricultural products. This fall in agricultural prices was due chiefly to the lack of ade-

---

[1] *Report of Royal Commission on Agriculture, Punjab Volume* (London, 1926), pp. 410-439.

quate transportation facilities to carry off the surplus produce in the years of good crops. This close dependence of agricultural advance upon transportation facilities seems to have been overlooked by certain writers on Indian economic problems who tend to criticise the policy of the British Government in extending the construction of railways in advance of a more rapid expansion of irrigation canals.[1] But the peasant in the newly irrigated areas, or in old areas more efficiently irrigated, must have some ready outlet for his excess produce if he is to make agriculture a profitable venture. A dual extension of railways and irrigation facilities was and still is economically necessary if the vast irrigation schemes projected are to prove successful in increasing the economic well-being of the people of the Punjab. It is evident that the immediate motive for the rapid extension of railway lines in India, and in the Punjab, by the British following the Mutiny of 1857-8 was the need for rapid means of transportation for strategic military operations. But that need not blind the student of Indian economic conditions to the economic function of efficient transportation facilities which the railroads have provided. By 1928 India could boast of over 40,000 miles of railway lines in efficient operation of which the Punjab accounted for 5,500 miles within its boundaries. The officials in the Punjab in 1861 expressed their recognition of the interdependence of agriculture and transportation in the following words:

Although the harvests on unirrigated lands were scanty throughout the Punjab, the yield from watered tracts was generally superior, both in quality and quantity. The great demand for corn in the Delhi territory and the extraordinarily high prices, led to its extensive exportation from the hills of Kangra and the wheatfields of the Jullundhur and Baree Doabs.

[1] Kale, V. G., *Introduction To The Study Of Indian Economics* (Poona, 1922), p. 283 *et seq.*

This trade has been especially beneficial to the owners of irrigated and inundated land, who have been much depressed by the cheapness of agricultural produce prevailing for so many years. The keenness with which exportation was carried on when prices became decidedly remunerative, proves that the people are fully awake to their own interests and that, to the extent of their means, they are as ready as other nations to avail themselves of any facilities for traffic which may be provided for them. The manner in which the Delhi territory, drained as it had been of food by war and famine, was supplied from Hindustan and the Punjab, is an illustration of the law of political economy which prohibits any interference of the state in the corn trade.[1]

Apparently the official who wrote the above paragraph into the report had in mind Adam Smith's remarks as to the desirability of non-intervention in the corn trade of a country:

In an extensive corn country, between all the different parts of which there is a free commerce and communication, the scarcity occasioned by the most unfavourable seasons can never be so great as to produce a famine; and the scantiest crop, if managed with frugality and economy, will maintain, through the year, the same number of people that are commonly fed in a more affluent manner by one of moderate plenty. The seasons most unfavourable to the crop are those of excessive drought or excessive rain.[2]

The point stressed in connection with the situation in the Punjab in 1860 was the relatively favorable position of that province compared to the territory lying about Delhi, making possible the export of wheat to the Delhi area by the *Punjabi* peasants in the irrigated sections in spite of lack of efficient

[1] *Administration Report*, 1860-61, p. 131.
[2] Smith, A., *Wealth Of Nations*, edited by Cannan, E. (London, 1925), 4th edition, vol. ii, p. 27.

means of transport. During these early years of British control of the Punjab, as indicated incidentally in an earlier section efforts were being directed not only toward the development of irrigation canals but also to the construction of roads and railroads. Eventually, the dread of famine in the Punjab was to be entirely removed through better means of producing food crops on irrigated lands and the development of an efficient system of transportation and communication.

By 1861 the Maharajah of Putiala was beginning to evince a keen interest in the possibility of developing an irrigation scheme which would irrigate the land in his state. He approached the British with the suggestion that he bear the expense of the preliminary survey and preparation of plans for the project under the direction of qualified British officers to be supplied by the British Government. This survey was immediately begun on the terms suggested. The following year another significant development was noted in the district of Dera Ghazee Khan on the frontier where a chief named Mussoo Khan had offered to excavate a canal on the condition that he be allowed to hold the land to be irrigated, free of revenue for a term of years. His example caused several other landed proprietors of the district to combine for a similar purpose, and the district officer had under consideration five separate offers from *zamindars* who wished to construct canals at their own expense.[1] This development, while not significant in relation to the total area which might be irrigated in such a manner, was important in showing that within eleven years of the annexation of the province, these frontiersmen were willing to cooperate with their British rulers in the development of irrigation schemes which would permanently enhance the stability of the means of livelihood of that section of the province. The Govern-

[1] *Administration Report*, 1861-62, p. 120.

ment sanctioned Mussoo Khan's project and at his own expense, he set himself to the task of digging a canal, which was to be 27 miles long. The fact that about half was completed within a year would indicate that this project probably called for the renovation of certain canal ruins in the area which he controlled. Mussoo Khan's canal was to derive its waters from the Indus River, taking off on the west side. Another *zamindar* had obtained permission from the Government to build a twelve-mile canal in the same district at his own expense.[1] Since the officials in charge of irrigation projects were seeking every possible means of increasing funds available for prosecuting the development of their canal schemes, this voluntary cooperation on the part of these chiefs was hailed with enthusiasm and was thus recorded in their report:

> The amount already spent by the chiefs, or immediately required of them, cannot fall short of a *lakh* of rupees, and the desire to contribute appears, as has been said, to be on the increase. There appears every prospect of the scheme, heretofore urged, of raising funds on the spot for carrying on irrigation works, by the issue of indentures or otherwise, being carried out on a limited scale amongst the native population of these parts. A decision, however, is still required as to the terms which are to be allowed and the extent and form of contract to be exercised, in regard to both which points the Financial Commissioner has hitherto delayed expressing a decided opinion, owing to the inadequacy of the existing establishment for the exercise of an effective supervision, and the great importance of adopting just principles at the outset in laying down a scheme of irrigation which will be wholly new in many essential particulars.[2]

However, as later events were to show, this source of

[1] *Administration Report*, 1862-63, paragraph 161.
[2] *Ibid., op. cit.*, p. 139.

funds for the prosecution of irrigation schemes in the province was not seriously considered in the years which followed. It never proved a promising source of funds for developing large irrigation schemes. During this period of the launching of new irrigation projects the funds for construction work were derived from Government grants out of revenue receipts. This policy continued until 1882 when, after years of adherence to the policy of paying for irrigation developments out of revenue as received, a plan was evolved which promised a more adequate fund for huge irrigation expansion. But during the sixties and seventies the engineers in charge of the construction work were forced to chafe at the delays occasionally caused by failure of the Government to grant sufficient funds for the most expeditious translation of plans into working projects. In spite of lack of sufficient funds and hindered occasionally by storms which destroyed the headworks in a state of partial completion, the work was doggedly pursued and each year witnessed additional channels completed and further areas irrigated.

A new scheme was seriously proposed toward the close of the year 1863 to irrigate the country lying between the Chenab and Ravi Rivers. A survey of this area was commenced and a preliminary study of the land to be affected led to the conclusion that the irrigation of the territory between the two rivers would be a profitable venture. It will be remembered that the Baree Doab Canal had taken off from the River Ravi and projected to the south from the east bank of the river. The new project was to irrigate a portion of the area to the west of the Ravi.

THE SIRHIND CANAL

In 1868 another large irrigation project was launched, namely, the Sirhind Canal. This canal was planned to take off from the Sutlej River so as to irrigate a wide stretch of territory between the Sutlej and the Jumna Rivers. It was also to act as a feeder channel to the irrigation system in the State of Putiala, references to which have been made previously. In 1870, the first financial return from this new project was entered in the Administration Report, indicating that Rs. 1,585 had been collected. The inclusion of this item in the report signified that the technique of canal construction had developed to the point where a portion of the new project was in operation and supplying water within a year or two of the initial launching of construction work. In 1871 a new item, referring to a scarcity of labor appears in the progress report. Heretofore it had been quite easy to get laborers in sufficient numbers to develop the work as rapidly as funds permitted. In the work connected with the Sirhind Canal construction it was necessary to try the experiment of employing prison labor on excavation work. This apparently proved quite successful and some four million cubic feet of earth were excavated by prison labor. About 500 prisoners were employed in such work in 1871 and it was decided to increase the number of prisoners so employed to 3000 the following year.[1]

As was indicated in an earlier section, one of the reasons for immediately launching construction work on new irrigation schemes in the Punjab so promptly after its annexation by the British was the problem of giving employment to the disbanded Sikh soldiers. The wage paid for unskilled labor in 1849 in this canal-construction work was about one and a half annas per day.[2] A few years later an increase in pay of a half anna per day was urged for the native soldiers in the Indian army, but this suggestion met with violent opposition on the ground that it would precipitate a similar rise in wages for unskilled laborers employed on Governmental construc-

[1] *Administration Report*, 1870-71, paragraph 237.
[2] *Ibid.*, 1849-50, pp. 41 *et seq.*

tion works. The work required of these unskilled laborers consisted chiefly of carrying excavated earth from the canal beds to the banks and this labor was performed by men, women and children who came from the agricultural villages in the areas through which the canal channels were being constructed. The work of the Indian peasant was so arranged, due to the seasons and the short period during which labor was required on the land, that he had many days to devote to any form of labor which might be offered. Hence, it is all the more surprising to find that in the construction of the newly launched Sirhind Canal laborers were scarce. There were two reasons for this scarcity of laborers: the area through which much of the new canal was being built was waste land quite far removed from settled villages; the other was that the villages near this section had been seriously affected by the famines of the years 1860 and 1868, almost to the point of depopulation.[1] The labor problem during the period of rapid expansion of irrigation projects in the Punjab, however, has never been very serious. In most cases the peasants have been eager to avail themselves of the opportunity afforded by the canal-construction work to add to their cash income.[2]

Another feature which is especially interesting in connection with the Sirhind Canal project is that this was the first of the British Canals to be financed by loans. Up to this time, as will be remembered from the previous discussion, canals had been constructed from grants made by the Government out of revenue receipts. After 1878 a new classification of irrigation works was adopted whereby certain canals were called "protective" and others were called

---

[1] *Cf. Administration Reports*, 1860-61 and 1868-69.

[2] Cash payments of wages to these unskilled laborers employed on canal construction works were in many cases the first introduction of money into the village economy.

"productive". The distinction was based upon the comparative earnings of the canals in respect to their capital costs. Thus, canals which were expected to be financially profitable in the sense of earning a fair return on the capital investment were to be called "productive" while those which promised to just about cover operating costs but no net return were to be called "protective works". In other words, the protective works were canals which helped protect the areas through which they were built against the possibility of famine, and thus made famine-relief measures in other forms unnecessary. The precise method of determining when, where and to what extent "protective works" might be developed was not settled until 1901 when the Indian Irrigation Commission recommended that:

In every district, in addition to the area already under irrigation, the areas which must be brought under irrigation with a view to protect the district against future famine is first estimated. In doing so, population, and the area which must be irrigated per head of population are taken into account. The latter item comes to about a third to half an acre. Nextly (*sic*) the annual average expenditure for famine relief in the future is estimated, after taking into account the cost of famine relief in the area in the past. The conclusion is that the state will be saved the estimated expenditure on famine relief, if the area as calculated above be brought under irrigation. In other words, if the interest charge on the capital outlay to bring the necessary areas under irrigation does not exceed this figure the Government will be in the same financial position as before.[1]

The capital expenditure which the Government might thus incur on this supposition was to be called, "indirect protective value of an irrigated acre". The Irrigation Commission at that time recommended that an expenditure up to

[1] Vakil, C. N., *Financial Developments In Modern India* (Bombay, 1925), p. 242.

a maximum of three times this indirect protective value as thus defined be made permissible. In addition to the protective works providing eventually insurance against famines the Commission indicated four indirect benefits which would accrue to the Government from developing "protective works": (1) the increase in the general wealth and prosperity of the tract thus brought under irrigation; (2) an increase in the humidity of the air of neighboring areas affecting the productiveness of those areas; (3) raising the level of the underground water supply, thus making wells less expensive to build and operate; (4) preventing the losses, demoralization and misery of famines over an area larger than that actually irrigated.[1] In a later section [2] the classification of all the canals of the Punjab as either productive or protective is indicated.

The point here stressed is merely the fact that the Sirhind canal in its later developments [3] was the first of the irrigation projects of the Punjab to be financed by loans floated in London. Politically, the Sirhind Canal raised certain practical problems of cooperation between the British and the native states through which certain branch canals, parts of the larger system, passed. In addition to Putiala State through which a portion of this system passed to the extent of supplying irrigation to 35% of the total area of land to be irrigated by the whole Sirhind Canal project, the Native States of Nabha, Jind, Faridkot, Malerkota and Kalsia were to benefit to the extent of 18% of the total area commanded by the project. The remaining 47% of the area commanded lay in British territory. *Apropos* of the problems thus raised is the following statement:

[1] *Indian Irrigation Commission Report, op. cit.*, vol. i, chap. iv.
[2] Chapter vii, *infra*.
[3] The Sirhind Canal system was not completed until 1887.

This policy of leaving large areas under native control must always create formidable obstacles to the projects of administrators anxious for progress. To take one instance, the lines traced by the engineers for canals and railways had no relation to the boundaries settled by history between native and British territories. In such matters reluctant princes had in the end to give way to the needs of the engineer. But administrators eager to push on their beneficent schemes naturally chafed at delays which had not to be faced in territories subject to direct British control.[1]

This probably accounts for the long-delayed completion of the Sirhind Canal project, though it must be stated on behalf of the Maharajah of Putiala State that as early as 1860, as has been indicated, he had requested the British officials to survey his territory for the purpose of bringing the benefits of irrigation into his State. There was delay, because of the necessity confronting these weak political units of devising means of financing that portion of the scheme which lay within their boundaries. The writer of the article in the *Round Table* perhaps had in mind the interference and difficulties which continue to this day to delay the construction of a huge irrigation scheme which the British have proposed in which the Woolar Lake lying entirely within Kashmir State is to provide the reservoir. As late as 1927 the necessary cooperation of the Maharajah of Kashmir had not been conceded on terms acceptable to the British.

PROPOSALS TO IRRIGATE THE LOWER BAREE DOAB

Another project which occupied the attention of the British engineers who were proposing various irrigation schemes during the decade 1870 to 1880 was that related to an extensive plan to irrigate approximately 11,000 square miles of territory in what is called the Lower Baree Doab.

---

[1] *The Round Table*, London, September, 1928, pp. 693 *et seq.*

It will be recalled that the Baree Doab Canal was the first great British irrigation canal projected in the Punjab. This canal was irrigating the area in the northern portion of the Baree Doab, that section of the Punjab lying between the Ravi River and the Beas River. Details as to the launching and development of this new project will be discussed in the next chapter. It is significant to note that the engineers estimated the probable return on the new canal in the lower portion of the *doab* at no less than 10½ per cent per annum on the proposed capital expenditure, thus bringing it easily within the scope of what were to be designated " productive works " for the construction of which funds might be borrowed outside of the Punjab.

Another extremely interesting proposal made in 1871 contemplated the construction of a complex system of canals in the territory named the Sind Sagar Doab lying between the Indus and the Jhelum and Chenab Rivers. However, this project was viewed as a decidedly impractical one at the time because of the scanty population of the tract to be irrigated. The land was well situated, however, for purposes of irrigation and the rivers on both sides of the tract offered abundant sources of available water for changing the entire aspect of this deserted territory. Other projects, however, offered less serious obstacles, and therefore this project was not seriously considered until some forty years later. When the scheme was finally developed, the original plans were used as a basis for projecting the most ambitious irrigation scheme thus far developed. In the chapter which follows the development of irrigation schemes in this waste land will be surveyed.

CHAPTER IV

Rapid Expansion of Irrigation and the Introduction of Colonization Schemes, 1880-1927

The Huslee Canal, the Western Jumna Canal and the Upper Baree Doab Canal which have been described in the previous sections of this monograph were built through sections of the Punjab which were quite thickly populated. The Sirhind Canal was projected through certain sections which were not so thickly populated but which nevertheless supported some settled village life. These four projects thus far reviewed were constructed with a view toward improving the position of the cultivators already occupying the territory through which the canals were to flow. Famine prevention, raising the standard of living and increasing the tax base were among the chief motives in the construction of these projects. Mention has been made of a proposal which urged the construction of irrigation works through the bars between the Jhelum, Chenab, Ravi and Sutlej Rivers. These areas, except for the low-lying portions nearest the rivers, were practically desert waste supporting no settled population. The low alluvial strips along the river bottoms were cut into shreds by many small inundation canals rudely constructed by the native population. Numerous villages, wells and green fields made these lowland tracts a striking contrast to the barren lands above the reaches of the inundation canals. Population and the prosperity of the countryside diminished rapidly as one moved inland and upward from the river banks. A few scattered wells in the drier regions immediately beyond the inundated area enabled a sparse scatter-

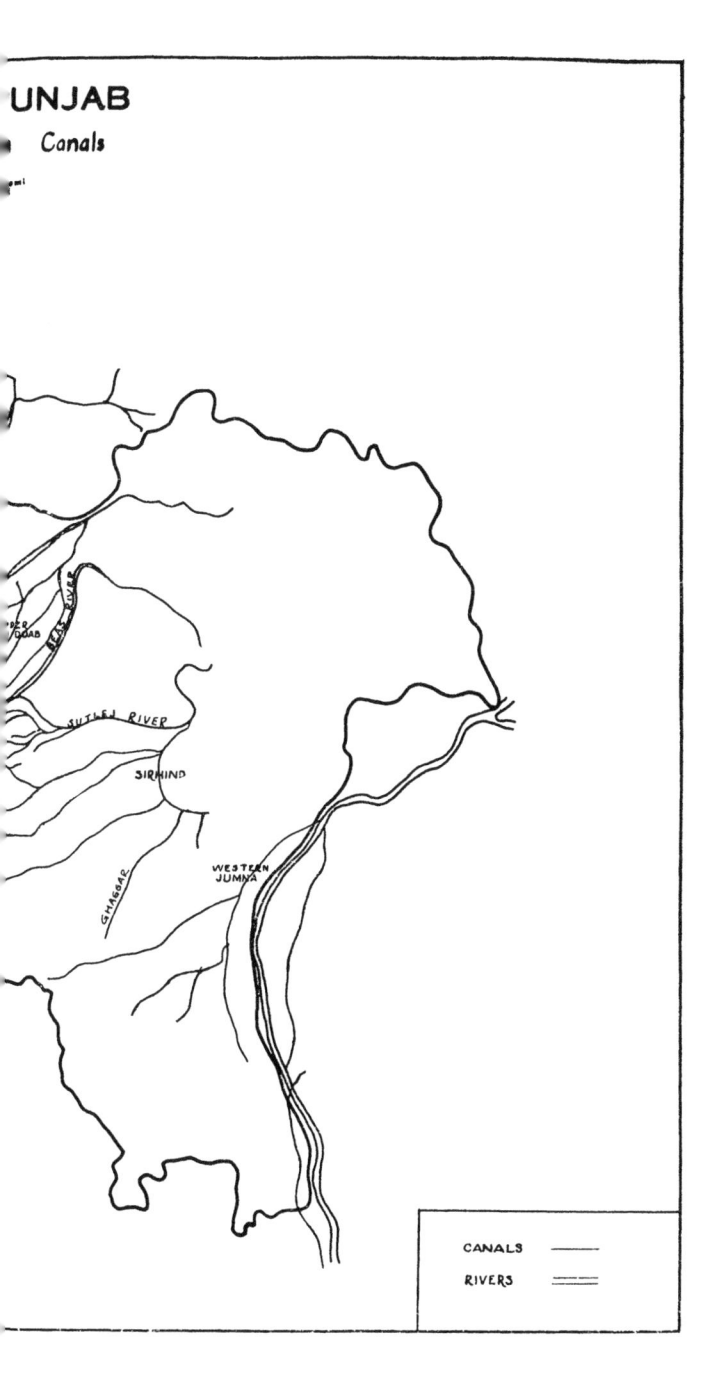

ing of hardy agriculturists to pursue their calling under quite unfavorable conditions. Beyond these last scattered cultivated patches the uplands presented a desolate waste. As late as the early nineties of the last century, according to Mr. Darling, a man journeying from the Jhelum to the Sutlej would have had to traverse 150 miles of some of the ugliest and dreariest country in the world.[1] Nothing but a few prickly shrubs, bushes and stunted trees grew on these bar lands. The area was inhabited by wolves, wild cats, jackals, and pigs while here and there were to be found dens of thieves who would raid the industrious population along the more fertile river lowlands. The rainfall in this tract was less than ten inches per year and the level of the subsoil water was as low as 50 to 80 feet, while water brought from that depth was brackish and unsuitable for irrigation purposes. Cultivation of the soil under such conditions was practically impossible.[2]

During the periods following the scanty rainfall these tracts would support a fairly abundant growth of grasses for a short season, during which the nomad tribesmen, known as *Janglis,* grazed their half-starved cattle over the area in search of pasture. These people made no permanent camps and apparently never cultivated the soil. They made a precarious living out of cattle-breeding and cattle-lifting. Camel thefts were not at all uncommon.[3] Occasionally the nomads would establish a random village settlement on the external fringes of the arable lands nearer the rivers, but they were unable to wrest a high standard of living from the natural resources at their disposal. The climate was subject to extremes of heat and cold, detracting still further from the

---

[1] Darling, *op. cit.,* p. 129.

[2] Puckle, F. H., *Punjab Colony Manual* (Lahore, 1926), vol. i, pp. 9 *et seq.*

[3] *Chenab Colony Gazetteer* (Lahore, 1904), pp. 24 *et seq.*

possible attractions of these niggardly tracts for settlement purposes. Strong winds blowing over the sand dunes shifted the contour of the rolling plains of sand with the passing of the seasons. The soil was loose and porous, consisting of a fertile loam where not too deeply buried under the sand. Much of the area would have been suitable for cultivation had it not lacked moisture. The lack of sufficient rainfall, the uncertainty of even the minimum amount arriving during the growing season and the uneven distribution of the rains over these desolate bar regions made necessary the development of irrigation schemes if a prosperous agricultural community were to be made possible in this area.

Thus the Government was confronted with a two-fold problem in projecting irrigation canals in this section of the Punjab, namely, the construction of the canals and the newer and more difficult problem of devising a colonization scheme to people the areas to be irrigated with a settled and industrious agricultural population. The engineering problems relating to the projection of the irrigation canals were quite simple on account of the abundance of water in the well placed rivers, the natural reservoir of winter waters stored up in the mountain snows, a soil which was easy to dig since stones and rocks were rarely found and the ideal lay of the gently southward sloping land. The remainder of this chapter is devoted to a brief survey of the canals constructed after 1880 and the successful colonization schemes by means of which the newly irrigated areas have been peopled with agriculturists.

LOWER SOHAG AND PARA CANAL PROJECT

The first canal-irrigated tracts to be colonized were those served by the Lower Sohag and Para Canals. These were two purely inundation canals taking off from the Sutlej River and irrigating a moderate area just east of the city of

Montgomery. The water supplied during the season from May to September was about 700 cubic feet per second, commanding a gross area of about 186,000 acres. One-half of this area was crown waste land which the Government divided into fifty-four village tracts. The chief interest in this relatively small project arises in relation to the problem of colonization of this waste land. The Government launched the project of choosing colonists and allotting land to prospective settlers in 1882 but the settling of the newly irrigated tracts was not completed until 1889. Colonists were chosen chiefly from among the agriculturists of the immediately adjacent areas so as to develop the minimum resistance to settlement in the new areas due to the peasant's innate repugnance to leaving his ancestral village. Details of the methods employed in colonization are discussed in a later section of this chapter. It is impossible to follow the financial progress of this small project since it was merged with the other inundation canals in the Upper Sutlej Inundation Canal System. On the whole, these inundation canals are termed "protective works" since they insure the area irrigated from famine, but no capital account is separately kept for this portion of the project. In good years when the flow in the rivers during the flood season is fairly steady and the level continues high enough to force the water through these inundation canals, the Lower Sohag irrigates approximately 80,000 acres.

THE SIDHNAI CANAL SYSTEM

The Sidhnai Canal was originally proposed in 1856 but it was not until 1882 that construction work was begun and it was finally opened for irrigation in 1886. It has its head on the Ravi River in the Multan district and now irrigates an area of over 200,000 acres in the western portion of the Ravi-Sutlej watershed. The Sidhnai system comprises four canals: the Sidhnai proper which is 37 miles long and the

Koranga, Fazal Shah and Abdul Hakim canals whose combined length is 31 miles. From these main canal channels some 384 miles of distributaries are fed.[1] The area commanded by this system of canals was 417,000 acres of which 232,000 acres were Government waste lands available for settlement.

Colonization of this large area was begun in 1886, the first year of the operation of the whole system. The success of the plan of colonization was doubtful for several years owing to difficulties which confronted the new settlers. The tract itself was unattractive jungle waste and the scattered indigenous population of the area was very unfriendly to the new settlers. So uninviting did the tract to be settled appear to the first group of would-be colonists that most of them refused to consider it as a profitable venture and returned to their villages in the more thickly populated portions of the province in preference to tackling the difficulties of tilling such an unfriendly soil. The Government was not slow in dispatching a hardier group of colonists selected from the densely populated portions of the Amritsar district. This second group apparently prospered from the moment of their arrival. Among the difficulties encountered by these early colonists in the Sidhnai canal area the most important were the clearing of the jungle of scrub bushes without adequate tools and machinery, leveling of the land thus cleared and the building of many small water courses through the fields. The policy pursued by the Government was to construct the main channels and branch canals but it was up to the settler to construct the water courses on his own plot for actual irrigation of particular areas. This called for the division of each acre into a minimum of ten squares into which, in succession, water could be led. This task proved rather arduous and in some cases it was found

[1] *Punjab Government Administration Report*, 1921-22, vol. i, p. 178.

that, due to faulty survey of levels, some of the plots were so situated as to make it impossible for flow irrigation to be utilized. To cultivate such plots required additional labor and capital in the form of Persian wheels, some form of lever pump, or coolie labor to carry the water from the canal levels to the land to be watered. Added to these difficulties were the continued raids and quarrels with the indigenous population which very often created disturbances far in excess of what might have been expected with reference to their numbers. These indigenous inhabitants of the bars were slow to see the advantage which would accrue from irrigation facilities to a settled agricultural community. Furthermore, these early settlers had no shelter. Everything had to be built from the soil itself and while wants were few and standards of living to which the colonists had been accustomed in their native villages moderate, difficulties were experienced in quickly providing even the most urgent necessities. Finally, there existed no easy means of communication and conveyance. Railways had not yet pierced that unproductive portion of the Punjab. The immigrant had literally to walk through a desolate country to get to the colonization area which proved to present a no less unpromising appearance than the wastes through which the land-hungry peasant had traveled. When, after severe hardship and toil the first crops were harvested and an excess of produce had been won from the soil, the surplus could not be sold to advantage due to the lack of markets and transportation facilities. Hence, even though immediate needs were fulfilled by the crops which were raised, agriculture on a profitable basis had still to await the development and construction of railways to the newly irrigated regions. While the capital required by the peasant was small, it was nevertheless available only at prohibitive rates of interest from money-lenders who soon followed the peasant colonists into the colony areas. The unfamiliar surroundings, the long days of heavy toil clearing the fields and preparing them for cultivation, the lack of appropriate implements, the complete isolation from their home villages due to the absence of efficient means of communication were factors which combined to make the lot of the peasant colonists in the early years of colonization difficult and burdensome.[1]

THE LOWER CHENAB CANAL

The Lower Chenab Canal which takes off from the Chenab River at Khanki was first proposed as an irrigation project in 1875. The original plans contemplated an enormous project with headworks above Merala to distribute water over the whole of the Chenab-Ravi watershed. Since this vast area was almost entirely waste lands except for the strips along the banks of the river, and therefore entirely lacking in an agricultural population, and since the Government had not yet had opportunity to experiment in colonization schemes in newly irrigated regions, the financial prospects appeared dubious and accordingly the project was not seriously considered as an immediate development. Thus, in spite of the tract's ideal situation for irrigation purposes, the absence of an immediate return on the capital investment reflected the Government's policy relating to new irrigation works. After the launching of the Sidhnai Canal System, which has been discussed, the British were unwilling to engage in any considerable irrigation developments which did not promise an early and profitable return on the capital costs of such works. Since the initiative for expansion of irrigation canals came entirely from the Government in power, it is difficult to present valid criticism of this profit motive in the policy relating to the expansion of canal irrigation.

[1] *Sidhnai Canal Completion Report* (Lahore, 1888), pp. 4 *et seq*. Cf. also *General Administration Report, Punjab Government*, 1888.

THE LOWER CHENAB CANAL 55

A smaller project, which was called the Ramnagar project, was thus proposed and plans sanctioned in 1884. It was to irrigate approximately 144,000 acres annually and its cost of construction was estimated at Rs. 30 *lakhs*. This canal was opened for irrigation in 1887 but it was a complete failure from the first. The estimates of its cost were greatly exceeded when the engineers' plans were executed and the prospects of its ever proving remunerative were extremely remote. The canal was constructed on an inundation basis and it was found that during the flood season the channels silted heavily and as soon as the river fell to its normal levels the silted channels were too high to permit the water to flow naturally from the river. There was no means of forcing the water into the channels to mature the crops which had been sown. In view of the uncertainty of the supply of water, colonization of the area was obviously an impossibility.[1] Hence, a new project was prepared and construction was begun in 1890. The new plan merely contemplated the enlargement of the system which had proved impractical to irrigate a total of 400,000 acres instead of the original 144,-000 acres. In the meantime, during the period of construction of the extensions the engineers in charge of the scheme prevailed on the Government to enlarge the scheme still further. In 1892 the Government authorized the construction of a larger canal with a head capacity of 8000 cubic feet of water per second to irrigate annually an area of 1,100,000 acres, at an estimated capital cost of Rs. 265 lakhs. The tract to be thus irrigated lay in the upper portion of the Chenab watershed, covering approximately one-half of the area which had been proposed as a possible irrigation project in 1875 as was stated in the introduction to this section. The only inhabitants of this tract were the indigenous nomads who eked out a precarious existence by means of their

[1] *General Administration Report*, 1921-22, *op. cit.*, vol. i, pp. 180 *et seq.*

camels and goats. They existed upon milk and cattle products. The country offered few attractions to the engineers destined to develop the project or to the laborers who were to spend several years aiding in its construction. By 1887 the Lower Chenab Canal was opened as an inundation canal and in 1892 it was completed as a perennial canal, the largest project of its kind which the British had constructed up to that time in the Punjab. Since practically the whole area commanded by this canal was waste land which prior to the projection of irrigation canals and distributary channels had not even been claimed as private property by the natives, this provided the British with a vast area for colonization purposes. Colonization began in 1892 and the colonists in the earlier years had an even harder time than usual.

Many of the colonists never reached the colony tract at all. Those who did found the tract inhabited by nomads who neither desired nor expected the canal to be a success and who were determined to do all in their power to prevent its being so. . . . A serious epidemic of cholera broke out and though those who survived the epidemic and had the pluck to persevere were rewarded by an excellent crop, their troubles were not yet at an end, as the labour available was insufficient to harvest it all, and, even when harvested, there was still the difficulty of selling the produce which had to go by the same perilous way by which the settlers had come. The opposition offered by the nomads was also a constant source of trouble and perpetual attacks were made by them on the colonists, who were, for some time, unable to ward them off.[1]

These hardships proved only temporary. By 1895 the soil had proved its fertility to the colonists and news of the good crops raised on the colony land quickly reached the ends of

---

[1] *Administration Report*, 1921-22, *op. cit.*, p. 184. The difficulties encountered by the colonists settling in the Lower Chenab Canal tract were very similar to those which have been described in the settlement of colonists in the Sidhnai Canal area.

the province. During the next five years colonists eager to settle in the Lower Chenab Canal area began a veritable exodus from certain of the more thickly populated portions of the province. Work on the construction of a railway line to serve this area was commenced in 1896 and by 1900 the line was opened providing safe and efficient transportation facilities from Wazirabad on the main line south through the Lower Chenab area to the city of Khanewal, connecting with the Lahore-Karachi main line. Thousands applied to the Government for land grants and it became possible for the officials to select the more desirable colonists for the new areas. The nomads also discovered that they were no match for the determined colonists and took advantage of the Government's offer of tracts on which to settle down to permanent agricultural pursuits. By 1922 these nomads had been quite successfully settled.

Their criminal ardour having been cooled by vigorous repressive measures, their disinclination to take land was gradually overcome. The belief that the canal had come to stay began to force itself upon them, and they found it advisable at least to make for themselves the best terms that they could. They were treated with great liberality in the matter of grants and have long since settled down to a peaceful agricultural life. They have acquired much knowledge from the colonists and most of them are now fair, and many of them decidedly good cultivators.[1]

The Lower Chenab Colony tract has proved to be one of the most profitable irrigation projects thus far constructed in India. Occupying portions of the Sheikhupura, Lyallpur and Jhang Districts, served by modern railway facilities and several good highways, this canal area is rapidly becoming one of the most progressive sections of the Punjab. Details as to its financial record are given in Chapter VII.

[1] *Administration Report*, 1921-22, *op. cit.*, p. 185.

During the decade 1894-1904 a colony was settled on the 100,000 acres of waste land situated at the lower end of the Upper Bari Doab Canal. This was called the Chunian Colony and its settlers were drawn almost entirely from the more congested portions of the Lahore District. About this time the Government was launching a scheme for relieving population congestion in certain sections of the province by inaugurating the policy of choosing prospective settlers chiefly from these thickly populated centres. Although the Chunian Colony is one of the smallest of the irrigation canal colonies of the Punjab it has proved to be a very successful venture as is shown in Chapter V in which the effect of irrigation extension on population shifts is studied. The Chunian Colony is also interesting because it was the first colony to be settled under the policy of lessening population pressure in a specific area.

THE LOWER JHELUM CANAL

Construction of the Lower Jhelum Canal was begun in 1888 but on account of Government's interest in the extension of the original Lower Chenab Canal project, as has been indicated, work on the Jhelum project was neglected. By 1897 the Government was ready to start work on the Lower Jhelum Canal with the result that it was formally opened in 1901. It was by no means completed by that date for many of its branches and distributaries needed still to be built. As construction on this project neared completion it became evident that the benefits of irrigation from this canal system could be greatly extended beyond the original scope of the proposal. In addition to a considerable extension of the actual construction work the new plans called for the inclusion of the existing Shahpur Inundation Canals in the Lower Jhelum System. The Shahpur Canals were among the indigenous inundation canals, badly in need of repair,

THE LOWER JHELUM CANAL 59

which the British found already constructed upon annexing the Punjab. During 1870-71 these canals had been renovated at considerable cost to the Government, without specifically determining the proprietary rights of the land owners occupying the land served by this system of inundation canals taking off from the Jhelum River. The inclusion of these canals in the Lower Jhelum system was dictated by the engineers' desire to include as large an area as profitably possible in the new project. Apparently the land-owners were not very carefully consulted as to the enlarged scope of the plan.

The project was vitually completed in 1908 when unexpected difficulties arose in connection with the Shahpur Branch. The withdrawal of water from the Jhelum had not proved nearly as injurious to the inundation irrigation as had been anticipated but the owners of the private canals in the tract, who were extremely jealous of their proprietary rights, refused to accept any terms of compensation which would permit the branch being run as a financial success. Government was naturally unwilling to press the owners against their will, to surrender their rights, and consequently the further construction of the branch was abandoned in 1916 and the money already expended on it accepted as a loss.[1]

This case relating to the Shahpur Inundation Canal appears to have been the most serious difficulty of this type encountered by the British in their relations with vested interests in land irrigated by older inundation canals which were eventually sought to be included in larger British works. The fact that the decision to suspend further construction of that portion of the Shahpur Branch in the largest scheme was not definitely reached until 1916, eight years after the difficulty was first seriously encountered, is evidence of an extremely long taxed patience on the part of the British with

[1] *Administration Report*, 1921-22, *op. cit.*, p. 187.

the interests affected. However, in spite of the fact that this portion of the Shahpur Branch had not been completed up to the end of 1927, the Lower Jhelum Canal system as a unit has proved a remunerative financial venture for the British as is indicated in Chapter VII.

The Lower Jhelum Canal has its headworks at Rasul on the Jhelum River. At that point a weir 4,100 feet long was constructed across the river to provide a discharge capacity for the main canal channel of 4,100 cubic feet per second down a channel which is 140 feet wide and 39 miles long. At the 39th mile the main canal divides into two branches, the Northern and Southern respectively, whose combined length is 208 miles. About 1000 miles of distributaries have been constructed in the Lower Jhelum System. The total area commanded by the Lower Jhelum Canal is 1,160,000 acres of which 568,000 were Government waste lands subject to colonization. Colonization in this area began in 1902 the year after the opening of the system. By 1921 439,000 acres of crown waste had been allotted to colonists. A large portion of the land thus allotted amounting to 240,000 acres, was cut by the Government into " service grants " for the special purpose of breeding horses for the army while some grants were made direct to the Army Remount Department for the same purpose. Details as to the basis of land grants are given in a later section of this chapter.

THE TRIPLE CANAL PROJECT [1]

One of the most recent irrigation schemes to be successfully completed in the Punjab is the so-called Triple Canal Project. It is so named because it is composed of three distinct units, the Upper Chenab Canal which was opened in 1912, the Lower Bari Doab Canal completed in 1913 and the

---

[1] *Cf.* map of Punjab showing railways and Triple Canal Project, frontispiece.

Upper Jhelum Canal first opened for service in 1915. While the project is thus made up of three separate canals, it is operated as a unit, utilizing and supplementing in its development the waters of the Jhelum, Chenab and Ravi Rivers. At Mangla the Upper Jhelum Canal carries the waters of the Jhelum River to the south-west, irrigating the land through which it courses, and then empties its unused waters into the Chenab River at Khanki. At this point in the Chenab River the Lower Chenab Canal takes off, carrying in its channel a composite of the waters of the Jhelum and Chenab Rivers. But the Lower Chenab Canal, as has been explained, was constructed earlier and is thus not a unit in the Triple Canal Project. The only connection that this canal has with the new project is that it now utilizes some of the surplus waters of the Jhelum River which have been emptied into the Chenab at Khanki by the Upper Jhelum Canal. The waters of the Chenab which are thus not required for the Lower Chenab Canal are freed for other irrigation purposes. Thus it has been possible to construct a new headworks at Merala, 36 miles above Khanki, where the Upper Chenab, the second link in the Triple Canal chain takes off. The Upper Chenab Canal irrigates portions of the Sialkot, Gujranwala, Sheikhupura and Lyallpur Districts but does not exhaust its waters during this process. Hence the Upper Chenab Canal crosses the Ravi River at Balloki and disposes of its remaining flow into another new canal which is an extension of itself but which is called the Lower Bari Doab Canal. This third link in the Triple Canal Project irrigates portions of the districts of Lahore, Montgomery and Multan. Thus, by means of this rather complicated engineering feat, an area which was once the most desolate portion of the Punjab has been converted into one of the province's chief cotton and wheat producing areas. The total area commanded by this huge project was originally estimated at 4,000,000 acres of which 1,675,000 acres were proposed to be irrigated annually. The actual irrigated area has considerably surpassed this original estimate which bids fair to insure an eventual profit from the scheme in excess of the amount originally estimated. Of the area commanded, 1,570,000 acres were subject to allotment or sale as Government waste land. In the Montgomery District a tract of 20,000 acres of Government land was sold at auction during March, 1915, and the average price paid was Rs. 275 per acre (about $100). The peasant farmers of the older canal colonies were eager bidders for this land but city dwellers belonging to the agricultural tribes also joined in bidding for land. The first crops on this auctioned land were reaped about two years later and up to 1921-22 the net return per acre could scarcely have been more than sufficient to pay a nominal rate of interest on the investment incurred by the peasant land owner. The peasant's return from using irrigated land is discussed in Chapter VII. In this Triple Canal Project the Government announced the following terms for disposal of waste land:

1. For peasant colonists of the older irrigated areas 680,000 acres were reserved, of which 175,000 acres were to be given to colonists on the basis of certain service conditions, such as, the breeding of camels or mares and the planting of tree nurseries. The remaining 505,000 acres were to be sold on the basis of eventual acquirement of proprietary rights following ten years occupation of the land in question. Payments for the land could extend over a period of thirty years.
2. To relieve congestion of population in certain areas of the Punjab 80,000 acres were to be allotted to peasants from such sections. Proprietary rights could be acquired after ten years' occupation of the soil and payment for the same could be extended over thirty years.
3. To be sold at auction, 125,000 acres.

THE COLONIZATION PROJECT

4. Reserved for agricultural and regimental farms, 100,000 acres.
5. 75,000 acres were to be reserved for the landed gentry who had served the Government conspicuously.
6. For irrigated forests and fuel reserve, 40,000 acres.
7. Allotments to semi-nomadic tribes that formerly pastured their cattle over the whole of this area, 40,000 acres.
8. For the amelioration of the 2,000,000 members of the depressed classes of the Punjab 30,000 acres to be reserved.
9. For awards to those who had served gallantly in the World War on behalf of the Empire 22,000 acres were to be reserved.

With slight changes in the details of certain of the items enumerated, the Government allotted the land as had been planned. By 1922 waste lands had been distributed in the following amounts in the Triple Canal Project area: in the Upper Jhelum Canal territory 43,000 acres; in the Upper Chenab Canal area, 85,000 acres; in the Lower Bari Doab Canal section, 1,442,000 acres.[1] It is thus evident that the Government was no longer finding it difficult to secure colonists for its newly irrigated sections.

THE COLONIZATION PROCESS

Before the actual construction of a canal is launched the tract to be irrigated is divided into large squares or rectangles for the purpose of determining the approximate position of the main line and branches in relation to potential colonies of agricutural settlers. Each of these divisions is a multiple of the smaller rectangle or square into which the land will later be divided for purposes of allotment. These large squares form the basis for the original survey of the canal system. The survey executed on the basis of the smaller divisions is used in determining the general layout of the distributaries and water courses. The whole tract to be

[1] *Administration Report*, 1921-22, *op. cit.*, vol. i, p. 194.

irrigated is thus cut up into equal and regular squares and rectangles, the size and shape of which may vary, however, in the different areas to be colonized. Thus, in the Sidhnai Colony area the squares were 22½ acres each. In the Lower Sohag, Lower Chenab and Lower Jhelum colonies the squares were 27.8 acres each. In the Triple Canal Colonies, Upper Chenab, Lower Bari Doab and Upper Jhelum, the rectangles were 25 acres each. Just what determined the differences in the size of the squares or rectangles is not indicated in any of the records which have been studied in the preparation of this monograph. It would be reasonable to suppose that the areas were different in size due to the varying fertility of the soil of the respective areas to be irrigated. However, on this point the records are silent and the writer was unable to discover the answer to his question from officials on the ground. The square or rectangle of 22½, 25, or 27½ acres, as the case might be, was the general unit of allotment. These units were then again subdivided into smaller squares of approximately an acre in extent. Finally, the cultivator cut up these acre plots into small sections of from one-tenth to one-twentieth of an acre in size for efficient control of water on the various portions of his field.

Thus the alignment of the water courses in the colony tracts preceded the actual creation of holdings. This made possible the fixing of the boundaries of each group of allotments to coincide with the areas most naturally commanded by the water courses which served them. For this purpose a contoured map which showed the natural drainage was prepared for each of the larger colonization areas. Villages were then plotted on the map in areas commanded by one or more of the larger water courses. It was so planned that no two villages should ordinarily share the same water course at risk to peace and order. After the village boundaries had been settled by the engineers, the main streets and general plan of the settlement were determined upon, keeping in mind the needs of the Punjab village in the matter of grazing grounds, accommodations for village servants and land to be devoted to communal purposes growing out of caste customs and traditions. Since all this was planned before the colonists arrived on the ground, it remained for them to build their houses and commence breaking up their land and building the water courses to bring the needed canal water to their particular fields. The colony villages which are thus methodically planned naturally possess marked sanitary advantages over the ordinary Punjab villages which just grow with the needs of the community without definite plan and certainly without serious attention to sanitary requirements. This task of laying out the canals, making maps, fixing village boundaries, laying out of streets and allotting areas for particular purposes within the village areas is a comparatively simple one which the engineer has been able to accomplish with great success. Land without laborers is, however, a poor basis for Government revenues. The problem of attracting settlers of the type to insure the success of the irrigation schemes in the waste regions of the province was a new one for the engineers and the officials in charge of the irrigation projects in the Punjab.

The selection of the individual colonist was generally left to the revenue officers of the districts from which they were drawn. These officers were in a position to know the applicants personally in addition to being on the ground and thus able to make inquiries into the record of every man who applied for land in the colony areas. As has been shown, the Government's motives in the projection of irrigation schemes in the Punjab were to increase the cultivated area of the province, to raise the standard of living of the people and thus widen the tax base for future revenue for the State, to prevent famines and to lessen the pressure of population upon the soil in the congested areas of the province. The importance of wise selection of the colonist was realized by the British for upon the colonist in the newly opened areas would the success of the scheme finally depend. The task of weeding out the ineligible from the mass of applicants was one of extreme difficulty. When once the profitable nature of agriculture in the newly opened areas became known, the people became land-hungry and were ready to try every conceivable scheme to gain an allotment of land. Among those rejected as ineligible for the privilege of becoming colonists were those whose families possessed sufficient land holdings, those whose older land holdings were mortgaged to a considerable extent, those who were physically or mentally unfit and those whose past record classified them as village loafers. When the unfit and ineligible had been eliminated there remained a group of men all connected by common ties since they had lived together in the same village or district, all physically fit to face the difficulties which would assuredly confront them in the newly irrigated areas, all short of land but otherwise solvent and with sufficient resources to attack the problems of a simple agriculture under new conditions. Such groups were sent to the colony areas in units about the size required to form the nucleus of a new village. Thus they started at the same time from the same village, taking with them the elements which go to form a separate village community. Knowing each other from their native village experiences in common, the stronger were thus in a position to aid the weak and the morale of the group was reinforced by the experiences of common hardships in the new villages. It is doubtful whether the colonies would have been successful had the officers in charge of the colonization scheme not decided upon sending the colonists in groups from the overpopulated villages and districts. The Punjab peasant would have been loath to migrate to the new territories alone to face the hardships and uncertainties of colony life among total strangers. Centuries of village life have moulded the attitudes of the individual member of such a village in the Punjab to an extent which practically prevents him from putting forth much serious effort and energy on his own initiative and responsibility. He is thus slow to cut the ties which bind him to his ancestral village. This aversion to change and breaking with village custom was lessened by the form in which colonization schemes were presented to the prospective settlers. Since the villagers could go to the new country in groups and then were allotted land sufficient to form the agricultural base for a new village community, the colonization plan was deprived of most of its terrifyingly adventurous aspects. Much of the success which has been attained in the canal colonies must be ascribed to the *esprit de corps* and the satisfactory communal philosophy of life which the colonists took with them to the newly irrigated territory.

The terms under which grants were made to the colonists varied considerably. The salient features were sufficiently similar however to permit a few generalizations. The grants appear to have been equal to one square in most instances, although grants of two squares were made to some especially favored colonists. Thus the allotment would amount to from $22\frac{1}{2}$ to $27\frac{1}{2}$ acres, depending upon the size of the squares or rectangles in the particular area. When it is realized that the average size of the holding in the province was very much less than $22\frac{1}{2}$ acres at the time the colonization scheme was first launched, it will be evident that the allotments must have appeared as remarkably large farms to these land-hungry settlers. In most cases the land was held on probation for a term of ten years by the colonists as tenants at will. In some of the earlier colonies inalienable occupancy rights were granted to the settlers at the end of this period of probation. These rights were granted sometimes free of charge and at other times on the payment of a nominal sum to the Government. In the later colonies, in the area served by the Triple Canal Project, for instance, a revised procedure has been introduced under which occupancy rights are granted after a first term of ten years and then, after a further term the tenants are given the option of purchasing alienable proprietary rights at a priviliged price payable on easy instalments.[1] It is evident that the colonization plans were evolved with the purpose of favoring the small land owner rather than the large. The Punjab thus continues to be a province of peasant proprietors to a greater extent, perhaps, than any other portion of India. The colonists can easily be classified into two large divisions, those receiving grants with service conditions and those receiving grants without service conditions. These two broad classes may again be subdivided into three groups: (1) peasant grantees; (2) *nazarana*-paying grantees; (3) military grantees. A brief explanation of these classifications will indicate their significance.

The major portion of the land grants in all the colonies was made to peasant grantees who thus make up the major element in the land-owning population of the colony areas. In the Lower Chenab Colony peasant grantees received 80 per cent of the land allotted; in the Lower Jhelum Colony 75 per cent of the land allotted went to peasant grantees; in the Lower Bari Doab Colony 59 per cent of the allotment went to the peasant grantees. These peasant grantees include both the imigrants from other portions of the province and the indigenous inhabitants of the irrigated areas. The imigrants were chosen chiefly from the congested areas of the

[1] *Administration Report, op. cit.*, pp. 174-75. If the colonist refuses to purchase alienable proprietary rights, he is permitted to continue the cultivation of his allotment as a tenant of the Government with occupancy rights.

Jullundhar, Hoshiarpur, Lahore, Gurdaspur and Amritsar districts. They were selected from among land owners or occupancy tenants possessing practical agricultural knowledge in the districts of their origin. The Deputy Commissioner of each district supplying colonists was instructed to select industrious, ambitious, skillful and enterprising cultivators for the colonies. Having been allotted a square or more of land in the new area, each peasant grantee was expected to prepare the soil for cultivation, build the channels for conducting the water from the distributary channels to his fields, develop his plot by careful industry, pay the land revenue, water rates and a yearly fee called *malikana* which was to be paid in recognition of the State's original proprietary title to the crown lands which were now being colonized. At the conclusion of a fixed period, generally ten years, the peasants having satisfied the original conditions of the grants, could on payment of a small sum secure inalienable proprietary rights as in the Lower Jhelum Colony, or alienable proprietary rights as in the Lower Chenab and Chunian Colonies.[1]

Nazarana-paying [2] grantees comprised (1) the yoemen and capitalists in the Chenab Colony, (2) civil grantees in the Jhelum and Chunian Colonies, (3) non-peasant grants to the landed gentry in the Lower Bari Doab Colony and (4) grants to individuals in return for services rendered to the Government. The yeomen grantees were descendants from agriculturists. They differed from the peasant grantees in being persons of more substance and of higher economic status than the latter. It was thought that the yoemen would cultivate their grants with the aid of tenants and that their credit resources might be made available for the development of the

[1] *Administration Report, op. cit.,* pp. 124-25-26.

[2] *Nazarana*—a due paid on succession to title; in this case it referred to an annual fee paid to the Government as a sign of the ownership of land which had previously been state property. Sometimes it takes the form of rent payment while at other times it is simply an additional tax payment.

colony areas. These hopes of the Government were ready fulfilled.

In every way these *nazarana*-paying grantees are unsatisfactory tenants of Government. Their endless disputes with tenants and among themselves, their migratory habits, retarded rather than helped the pace of the colony development. Similarly capitalists and the well-deserving servants of the Government did not come up to the mark. Their absenteeism falsified the expectation that they would serve as leaders of the new society.[1]

In other colonies yeomen grantees similarly failed to come up to the Government's expectations as to leadership and this type of grant was discontinued after 1901 largely due to the failure of the yeomen grantees of the Lower Chenab Colony to measure up to expectations. Civil grantees in the Jhelum and Chunian colonies were chiefly persons rewarded for services to the Government. Details as to the nature of the grants, the conditions under which they were made and the specific dues which were payable to the Government are not given in the available reports. Nor is the basis on which the fee was levied indicated by the available source material. Military grantees received allotments due to military service. In most cases their allotments were of the same size and were subject to the same conditions as the peasant grants in the matter of payment of water rates, Government dues, cultivation of the soil and a period of probation. However, the military grantees were not expected to pay the Government for their land after the period of probation was completed. Some grants of this type have also been made to pensioners of the pioneer native regiments of the Indian Army.

Grants with service conditions have been made in the

[1] Young, P., *Report On Colonization Of The Rakh And Mirnali Branches Of The Lower Chenab Canal* (Lahore, 1905).

Lower Chenab, Lower Jhelum and Lower Bari Doab Canal colonies. Among the service conditions imposed was the maintenance of camels for transport purposes. These grantees were generally given one or two squares of land on condition that they maintained a camel or two per square. These camels were needed by the British Army in the Punjab. Camel-maintenance grants have not proved very successful and it is doubtful whether this type of grant will be continued in the future. Other service grants relate to maintenance of mares for breeding purposes. Such grants have been made in the Lower Chenab Colony, Lower Jhelum Colony and in the Lower Bari Doab Colony. The allotments on these service conditions have proved successful to a high degree in the Lower Jhelum and Lower Bari Doab colonies where the area thus given was not less than two squares per service grantee. The demand of the Indian Army for horses and mules provides a steady market for the animals raised by the grantees. Another service grant which has been extensively used is that connected with the planting of tree nurseries. These grants have also been successful in many instances, as is indicated by the development of extensive groves of trees along many of the important canals. One of the important products of these groves will be fuel which, it is hoped, may make possible the utilization of the farmyard manure for fertilizing purposes in the future.

The camel-raising grantees were mainly *Biloches* of the Sandal Bar in the Lower Bari Doab Canal colony. They were expected to maintain one camel per square, or in some cases, per half square. They were required to register one attendant for every three camels. The land thus granted on service conditions in this colony could not be divided among the heirs of the grantee and hence the local government was called upon to select an heir on the death of the grantee. This practice led to the practical adoption of the rule of primogeniture in the case of the camel-breeding and horse-breeding grantees in the Lower Jhelum colony. This rule of primogeniture is by no means acceptable to the Hindu colonists whose traditional law of succession demands an equal distribution of the wealth of the father among the sons. In the Punjab the number of sons is sometimes fairly large and this leads to extreme subdivision of holdings. The Government attempted to forestall the tendency toward such minute division of the land by preventing the division of the land granted on service conditions among the surviving male heirs. However, owing to political difficulties in the Lower Bari Doab Colony, this rule has not been applied to service grants in that colony. In Chapters V and VI the problem of subdivision of holdings in the canal-irrigated areas is discussed in some detail.

We have thus far reviewed in broadly descriptive terms the construction of the chief productive irrigation works in the Punjab. The attempt has been to present the development of these projects in chronological order. In the order of their opening the canals which have been discussed are the Western Jumna, Upper Bari Doab, Sirhind, Sidhnai, Lower Chenab, Lower Jhelum, Upper Chenab, Lower Bari Doab and Upper Jhelum. In addition to these productive works in operation a huge new work is now under construction under the name of the Sutlej Valley Project. This work was commenced in 1922. When completed it will take its waters from the Sutlej River and irrigate the arid country comprising the native states of Bahawalpur and Bikaner as well as parts of the Nili Bar country in the Montgomery and Multan districts. The aim of this project is to insure from May to September a regular water supply to the area to be irrigated which is now served by inundation canals providing insufficient water. It is furthermore planned to increase the area now served by the existing inundation canals and finally the project when completed is to provide perennial irrigation to the uplands on both banks of the Sutlej River. This will eventually add two million acres of perennially irrigated land to the cultivated area of the province and three million acres of non-perennial irrigation. This project is being constructed with modern trenching machinery to a degree never before used in the Punjab. The use of modern methods of construction is hastening the project's completion in a manner quite new to India. This project will include the Gang Canal, Grey Canal, Dipalpur Canal, (all taking off from the Ferozepore Weir), the Pakpattan Canal (completed and in operation in 1926-27 irrigating 26,256 acres during its first year), the Eastern Sadiqi Canal, the Forwah Canal (the last three canals taking off from the Suleimanke Weir), the Mailsi, Bahawal and Quainpur canals (taking off from the Islam Weir) and a non-perennial Bahawalpur Canal taking off from the Panjnad Weir. This is the most ambitious project thus far launched in the irrigation program of the British in the Punjab.

New projects [1] not yet under construction but planned and awaiting development include: the Haveli project destined to irrigate portions of the Jhang and Multan districts; the Thal Canal which is to irrigate the Sind Sagar Doab lying between the Indus and the Jhelum rivers; certain dams to form reservoirs in the hills to enable the more economic utilization of the summer waters which still go through the river channels of the Punjab unused are proposed among which is the Bhakra Dam Project destined to irrigate the dry area between the Sutlej and the Jumna rivers. The latter scheme makes provision for a large modern hydro-electric development in addition to the irrigation proposed. A project which has been under consideration for many years and which has been cursorily mentioned in an earlier section in the Woolar Lake Barrage scheme. But the lake which is to form the

[1] *Cf.* appendix G. showing areas to be irrigated by new irrigation projects.

reservoir of this development lies within the confines of Kashmir State and the Maharajah of Kashmir has apparently not been convinced to date as to the advantages which would accrue to his State from cooperation with the British in this development.

In addition to the canals in operation which have been referred to, there is a series of inundation canals including the Chenab Inundation Canals and the Muzzafargarh Inundation canals. These two inundation canals have during the past few years averaged a combined area of 523,884 acres irrigated annually. Other inundation canals which are classed as unproductive works are those along the Indus River, the Shahpur Inundation Canals and the Ghaggar Canals. It will be remembered that the distinction between the productive and unproductive canals is based upon their financial record. Those canals whose yearly revenue indicates a fair return on their capital costs are termed productive; canals which fail to earn a fair return upon capital costs are termed unproductive. These unproductive canals averaged during the three years, 1924-5 to 1926-27, no less than 335,437 acres of irrigation service. Finally, the Lower Sutlej Inundation Canals for which no capital accounts are kept, irrigated another 349,768 acres annually as an average from 1924 to 1927.

Thus in 1926-27, the last year for which official figures are available, Government-operated canals in the Punjab irrigated 11,157,624 acres of land (approximately 17,434 square miles). Well irrigation added to the canal irrigation brings the total area under irrigation at the end of 1927 to 14.2 million acres. The total cultivated area of the province is about 30,000,000 acres, so that nearly one-half of the cultivated area of the Punjab is under either well or canal irrigation.

In the chapters which follow the economic significance of certain of the factors relating to irrigation in the Punjab will be studied.

PART II

SOME ECONOMIC ASPECTS OF CANAL IRRIGATION IN THE PUNJAB

CHAPTER V

THE EFFECT OF IRRIGATION ON POPULATION

ACCORDING to the Census Report of 1868 the population of the Punjab was 16,250,000, while the total cultivated area of the province was 20,172,000 acres. Of this area the portion irrigated by existing canals was 1,373,000 acres, while native wells irrigated 4,612,000 acres, leaving 14,187,000 acres under unirrigated cultivation. The cultivated area in 1868 amounted to about 1.25 acres per capita while the canal-irrigated area at that time was approximately .06 of an acre per capita. It is evident that most of the cultivated acrea was dependent upon the rainfall for its crops. When it is remembered that eighteen years of British rule had been occupied with the development of irrigation projects such as the Bari Doab Canal, the renovation of the Western Jumna canals and certain of the inundation canals along the Indus and Sutlej rivers, and that the irrigated area had been considerably enlarged by 1868, the dependence of the population on unirrigated land for its subsistence crops must have been considerably more severe prior to the British occupation of the Punjab. It is not surprising that famines were frequent in this area of India during the pre-irrigation era as was indicated in Chapter II. It also seems probable that an acre and a quarter of land, three-quarters of which were unirrigated, would have provided but a scant produce for the support of the individual thus dependent upon a variable income from the land. As has also been indicated in a previous section the lack of transportation facilities lessened the likelihood of the relatively less scanty produce of the poorer years reaching those portions of the province most in need of food to tide over the famine periods.

The population of the Punjab had increased to 17,270,000 by 1881 representing an increase of 6.3 per cent over that of 1868. The cultivated area in 1881 was 23,400,000 acres representing an increase over the 1868 figures of 16 per cent. This increase in the cultivated area was chiefly due to the enlargement of the Western Jumna Canal and the completion of the new Bari Doab Canal in 1873 and the partial opening to irrigation of the tract irrigated by the Sirhind Canal which was under construction at the time the Census Report of 1881 was issued. A definite sign of progress in 1881 was the fact that the rate of growth in the cultivated area was more rapid than the rate of population increase. Population while it was increasing was not pressing upon the means of subsistence, measured in terms of acres per capita under cultivation, as seriously in 1881 as it had been in 1868 and the earlier period of British occupation of the Punjab.

The Census Report of 1891 showed that population had increased during the previous decade by 9.6 per cent and that there were then 19,000,000 people in the province. The cultivated area had increased 11 per cent during the decade amounting to 26,000,000 acres, thus continuing the relatively greater increase in cultivated area compared with the rate of population increase. The increase in area cultivated was chiefly due to the completion of the extensions on the Western Jumna, Sirhind and Upper Bari Doab Canals during the decade. With the enlargement of the irrigated area lands lying just beyond the actually canal-watered areas tended to be brought under cultivation also, thus increasing the area under the plow by an amount larger than the increase in area irrigated. This is due to some extent to the gradual rise of the subsoil water-level when irrigation of a tract begins, thus making it possible to irrigate by means of wells on the nearby tracts. While there is a considerable time lag between the initial irrigation by means of the canals of a given tract and the gradual enlargement of the area under cultivation due to the rising of the subsoil waters, such extension occurs. Even though wells are not constructed and the land lying next to the irrigated areas is brought under cultivation the yield per acre on such newly worked land tends to be higher than other un-irrigated land yields at a distance from the irrigated sections.

In 1901 the Census Report gave the Punjab a population of 20,330,000 representing a 6.7 per cent increase during the decade 1891-1901. The cultivated area during the same decade increased by just about 7 per cent, bringing the total to 27,800,000 acres. Again the cultivated area increased slightly more rapidly than the population pressure on the soil advanced. In fact, from 1868 to 1901 the pressure on the soil was actually becoming less when the province as a whole is considered.

The situation in 1911 shows a 1.8 per cent decrease in the population of the Punjab as only 19,975,000 people were enumerated; the cultivated area by 1911 had reached 28,300,000 acres due to the rapidly expanding program of canal irrigation of the decade 1901-1911. Thus the cultivated area increased during the decade approximately 1.8 per cent over the area in 1901.

The 1921 Census gave the population of the Punjab as 20,685,024 and the cultivated area as 29,000,000 acres. The situation as it developed between 1868 and 1921 is summarized in Table I.

Between 1868 and 1921 the population of the Punjab increased by 27 per cent from 16,250,000 to 20,685,024 while the cultivated area increased from 20,172,000 acres to 29,000,000 acres, an increase during the period under re-

TABLE I. POPULATION IN RELATION TO CULTIVATED AREA AND CANAL IRRIGATED AREA IN PUNJAB, 1868-1921 [1]

Year	Population	Increase or decrease in decade	Cultivated area	Increase in decade	Acreage under canal irrigation	Percentage of canal irrigation to cultivated area
1868..	16,250,000	..	20,172,000	..	1,373,000	6.3%
1881..	17,270,000	6.3%	23,400,000	16.0%	1,950,000	8.3%
1891..	19,000,000	9.6%	26,000,000	11.0%	3,016,456	11.6%
1901..	20,330,000	6.7%	27,800,000	7.0%	5,000,551	18.0%
1911..	19,975,000	—1.8%	28,300,000	1.9%	7,227,042	25.0%
1921..	20,685,024	2.1%	29,000,000	2.4%	10,273,681	36.0%

view of no less than 44 per cent. The cultivated area per capita in the Punjab in 1868 was 1.25 acres. In 1921 it had risen to 1.41 acres per capita. However, even these figures do not indicate the full significance of the changed economic status of the province in relation to the cultivated area. An even more important point of difference is that relating to the relative portion of the cultivated area which was under canal irrigation in the beginning and the end of the period. Thus, in 1868 about 6.3 per cent of the area under cultivation was served by canal irrigation while in 1921 no less than 36 per cent of the cultivated area was served by irrigation canals. By 1926-27 the cultivated area had risen to 30,000,000 acres, the area served by irrigation canals was 11,157,624 acres or approximately 17,434 square miles of canal-irrigated soil. Thus it is evident that the extension of the cultivated area, largely due to the development of canal irrigation works since 1868 has greatly extended the base upon which food may be produced for the people of the

[1] Details taken from Census Reports, 1868, 1881, 1891, 1901, 1911, 1921. In 1926-27, the area under cultivation was 30,000,000 acres, the area served by canal irrigation, 11,157,624 acres. The percentage of canal irrigated area to cultivated area in 1926-27 had risen to 37 per cent.

province. Since population has not advanced as rapidly as the cultivated area has increased, it is evident that the per capita food-producing base was considerably larger in 1921 than it was in 1868, making possible a considerably enhanced standard of living measured in food-producing capacity per capita of the population. And finally, the irrigated area in 1921 was 36 per cent of the total area under cultivation as compared with 6.3 per cent of the cultivated area served by irrigation canals in 1868. Thus the uncertainty of the food supply of the province has been greatly reduced. The comparative productivity of the canal-irrigated land and the non-irrigated areas is discussed in Chapter VI. It seems reasonable to conclude from the analysis of these figures relating to canal irrigation and the enlargement of the cultivated area of the Punjab since 1868 that the people of the province are better fed now than they were in 1868 or 1849 when the British became interested in the economic position of the Punjab. This conclusion is supported by Mr. H. Calvert [1] who summarized the situation as follows:

It may be said that population has increased by about 20 per cent while cultivation has increased 50 per cent, and the gross value of the produce has risen from roughly 35 *crores* of rupees per year to not less than 100 *crores*.

It will be remembered from the discussion in Chapter IV that the Government hoped to relieve the pressure on the

[1] Calvert, *op. cit.*, p. 68. Mr. H. C. Calvert, a member of the Indian Civil Service, has for many years been a careful student of economic problems in the Punjab. His appointment to the post of Registrar of Cooperative Societies by the Punjab Government placed him in a peculiarly strategic position to study the economic condition of the province. His years of sympathetic scholarly study have made him one of the influential and well qualified consultants on economic conditions of the Punjab. His book, *The wealth and welfare of the Punjab*, has been one of the best source books for the study of the peculiar economic position of the Punjab since 1922.

soil in some of the more densely populated sections of the Punjab by means of its irrigation projects. Colonization schemes were not seriously developed until 1882 since irrigation development prior to that date was chiefly centered in areas already supplied with a numerous population. Hence the statistics relating to the density of population in the vari-

TABLE 2. DENSITY OF POPULATION PER SQUARE MILE OF THE VARIOUS DISTRICTS OF THE PUNJAB [1]

Districts	1881	1891	1900	1911	1921
Jullundhar	552	634	641	560	575
Amritsar	561	623	643	553	583
Ludhiana	426	447	464	262	391
Gujranwala	251	286	320	262	270
Lahore	280	334	374	372	420
Sialkot	524	572	555	518	522
Gurdaspur	436	500	498	443	451
Hoshiarpur	401	450	440	409	413
Ambala	442	459	434	367	362
Gujrat	284	315	309	307	322
Ferozepur	174	207	223	224	256
Rawalpindi	233	264	276	271	281
Jhelum	147.5	152.4	148.7	184	172
Multan	94	107	120	137	150
Shahpur	86	107	109	144	161
Jhang	113	117	123	152	165
Montgomery	78	94	97	108	154
Lyallpur	12	9	181	261	301
Sheikhupura	136	158	203	206	247
For all Punjab	152	167	178	174	183
Average density of colony district areas	148.2	169.6	204	216.5	243.3

ous districts of the province from 1881 to 1921 will show the degree of effectiveness of the plan to relieve population pressure by selecting colonists from the thickly populated areas. Table 2 shows the relative changes in population density.

Table 2 shows that there had been considerable fluctuation

[1] Data for this table taken from 1921 Census Report for the Punjab.

IRRIGATION AND POPULATION

in population as measured by density per square mile in the various districts between 1881 and 1921. The maximum density per square mile was reached in 1891 in the four submontane districts of Sialkot 572, Gurdaspur 500, Hoshiarpur 450 and Ambala 459. By 1901 the densities for the four districts had fallen to 555, 498, 440 and 434 per square mile respectively. While it is impossible to indicate in detail the movement of population from district to district, it appears from Table 4 *infra* that the drop in the density of population of the four districts was at least partially accounted for in 1901 by the emigration from these districts of colonists to the Lower Chenab Colony where in 1901 103,390 settlers indicated their district of origin as Sialkot, 43,593 had come from Gurdaspur District, 35,099 from the Hoshiarpur District and 7,777 from the Ambala District. Other areas were being colonized during this period and they probably accounted for a portion of the decline in population density per square mile in the districts in question in the Census Report of 1901. Two submontane districts which in 1901 reached their maximum density of population per square mile, Jullundhar District 641, and Ludhiana District 464, showed a decline in population per square mile in 1911 to 560 and 262 respectively. In 1911, as shown in Table 4 *infra*, colonists in the Lower Chenab Colony from the Jullundhar District numbered 70,847 and from the Ludhiana District 28,306. The Amritsar District reached its maximum density of population per square mile in 1901 when it was 643 per square mile. This fell to 553 in 1911. In that year Amritsar District colonists in the Lower Chenab Colony as shown in Table 4 *infra* numbered 81,144. By 1921, however, the population of the Amritsar District had reached 583 per square mile and if that increase is continued, the 1931 Census will probably show a density as high as it was in 1901. In the Lahore District, in spite of considerable numbers of colonists sent from that section to the various colony areas, the density per square mile has shown a steady tendency to increase through the years, with the exception of the slight fall from 374 per square mile in 1901 to 372 in 1911. Thus the density of population per square mile in the Lahore District in 1881 was 280 while in 1921 it had reached 420. For the province as a whole, the average density of population per square mile in 1881 was 152 reaching 183 per square mile in 1921. The average density of population in the colony areas in 1881 according to Table 2 was 148.2 while in 1921 it was 243.3. This shows that even though the population of the province per square mile increased during the forty-year period under review the less densely settled areas increased most rapidly in population, thus tending to develop a rough equalization of population density through the province. It is important to remember, however, that the area of the Punjab as a whole is not as important a criterion for judging the economic conditions of its people as is cultivated area per capita. Thus, while population increased from 152 to 183 per square mile from 1881 to 1921 the cultivated area increased from 23,400,000 acres to 29,000,000 acres as indicated previously in this chapter. At the time of the first British Census of the Punjab in 1868 the cultivated area per capita was 1.25 acres. In later years as indicated by Census Reports the cultivated area per capita varied as follows: 1881, 1.35 acres; 1891, 1.37 acres; 1901, 1.37 acres; 1911, 1.41 acres and in 1921, 1.42 acres. It is thus evident that there was a larger cultivated area per capita in 1921 than at any time since the British gained possession of the Punjab.

As was shown in Chapter IV the Sidhnai Canal, irrigating some 200,000 acres in the Multan District, was completed in 1886 and colonization began in that year. The statistics relating to immigrants into the Multan District as returned by the 1891 and 1901 Census Reports are indicated in Table 3.

TABLE 3.[1] IMMIGRANTS INTO MULTAN ON ACCOUNT OF THE SIDHNAI CANAL

District of origin	1891	1901
Amritsar	2,226	3,854
Sialkot	1,528	2,451
Lahore	3,677	5,099
Gujranwala	1,502	1,616
Jhang	24,751	25,439

This Sidhnai Canal Colony did not offer much relief to the congestion of population in the more thickly populated areas. The colony itself occupied a relatively small area of newly irrigated land and the area was fairly well populated by local tenants prior to colonization. The neighboring district of Jhang furnished most of the immigrants to this colony. But even in the case of the Jhang District, in spite of the loss of 24,751 emigrants who settled in the Multan District between 1881 and 1891, the pressure on the Jhang area increased from 113 per square mile in 1881 to 117 in 1891. At best, apparently, the opening-up of areas for colonists aided the Jhang District merely in retarding what would probably have proved a greater increase in population than has occurred. The population density per square mile in the Jhang District has increased steadily from 113 in 1881, to 117, 123, 152 and 165 in 1891, 1901, 1911 and 1921 respectively.[2]

With the opening of the Lower Chenab Colony tract of 2,500,000 acres to irrigation and colonization some immediate relief for the congested districts was possible. Colonization of the tract began in 1893 and by 1901 no less than 442,445 immigrants had settled in this colony area. This number was increased to 579,304 in 1911 as Table 4 indicates.

[1] *Multan District Gazetteer*, 1903-1904, quoting from the Census Reports of the Punjab.
[2] Table 2, *supra*.

TABLE 4. IMMIGRANTS INTO THE LOWER CHENAB COLONY AREA[1]

District of origin	1901	1911
Sialkot	103,390	96,984
Amritsar	67,963	81,144
Jullundhar	56,983	70,847
Gurdaspur	43,593	52,701
Hoshiarpur	35,099	44,234
Lahore	28,620	28,176
Gujrat	25,352	25,174
Ludhiana	17,807	28,306
Shahpur	16,156	12,367
Ferozepur	15,048	10,813
Ambala	8,614	17,242
Multan	7,777	12,671
Patiala	4,281	8,324
Jhelum	4,224	[2]
Kapurthala	3,968	8,129
Hissar	1,834	[2]
Rawalpindi	1,736	[2]
Mianwa'i	....	5,856
Montgomery	....	68,581
Other provinces	....	7,755
	442,445	579,304

Table 4 reveals the fact that there was a considerable degree of mobility among the colonists who apparently were not all willing to become long-time settlers. Thus, while in 1901 the immigrants from the Sialkot District to the Lower Chenab Colony numbered 103,390, ten years later those giving Sialkot District as their district of origin had fallen to 96,984. This would seem to show that quite a number of colonists failed to remain on their allotments of land long enough to obtain the rights which accrued to those who fulfilled the conditions the Government laid down when the land was granted. A similar shrinkage in numbers stating their district of origin is evident in the case of colonists from the

[1] *Chenab Colony Report* 1901 and 1911.
[2] Figures not available.

districts of Lahore, Shahpur and Ferozepur. There appears to have been no concerted action on the part of relatively large groups to return to their original villages. This failure on the part of some to take advantage of the opportunities afforded in the new canal colonies to become land owners is probably to be explained on the ground of individual inability of adjustment to the needs and conditions of the new community life. The colonization of the Lower Chenab Canal tract made possible a considerable shift of inhabitants from the more thickly settled districts of the province. The net effect of such shifts on the relative density of population per square mile in the various districts by 1921 hardly offers ground for optimism regarding this part of Government's policy in relation to irrigation. Table 2 *supra* would tend to show that the temporary shift of population makes room for additional members in the communities which remain in the older regions, with the result that after a relatively short period, the effects of the diminution of the density gradually tend to be erased. If such proves to be the case, the shifting of population will be but a temporary postponement of the serious problem of population pressure upon the available food-production base. It must be borne in mind, however, that the important aspect of the problem is not density of population per square mile of area, but cultivated area per capita available for production of food. This point will be stressed in Chapter VI. By extending the cultivated area steadily at a rate slightly in excess of the increase in population, the Government has been able thus far to lessen pressure upon the soil to some extent.

Two colonies which accounted for allotments amounting to 568,000 acres and 320,000 acres respectively, were the Lower Jhelum Canal Colony and the Upper Jhelum Colony. The former had attracted approximately 140,000 colonists by 1911 chiefly drawn from the districts of Gujrat, Sialkot,

Jhang, Jhelum, Gujranwala and Amritsar. More than 30,000 of the colonists came from the Gujrat District which borders the Lower Jhelum Colony tract. Not only was it easier to transport colonists in fairly large numbers from the Gujrat District on account of its proximity to the area to be colonized but of more importance to the success of the new colonization scheme was the fact that the population of the Gujrat District was Mohammedan. The Lower Jhelum tract lay chiefly in the vicinity of Muslim people and the inhabitants of the area prior to its colonization were followers of Mohammed. To prevent friction as much as possible it was deemed necessary to settle this tract chiefly with Muslim colonists. This difficulty of religious communal allegiance is one which confronted the British in practically every new colony founded in the newly irrigated areas. The strongly developed religious group-consciousness of the people of the Punjab has divided the population into opposing groups whose mutual suspicion creates a difficult administrative problem for the British Government.

The second of the two colonies just mentioned was a part of the Triple Canal Project which has been described. The Upper Jhelum Canal, being chiefly a feeder channel, carrying the waters of the Jhelum River across the Gujrat District to the Chenab River where it is immediately emptied again into the Lower Chenab Canal, offered but incidental colonization possibilities. What little land was available for allotment was divided among the disappointed would-be colonists of the Lower Jhelum Colony who had been chosen chiefly from the Gujrat District. Little relief was provided by the Upper Jhelum Canal to population congestion.

The other two canals in the Triple Canal Project, the Upper Chenab and the Lower Bari Doab, offered more extensive colonization possibilities. The former was opened for irrigation in 1912, the latter in the following year. As shown in

Table 5, these two canals made possible a shift of no less than 152,428 people from the more densely populated districts of the Punjab between 1911 and 1921.

TABLE 5. NET IMMIGRATION TO THE UPPER CHENAB CANAL COLONY AND TO THE LOWER BARI DOAB CANAL COLONY DURING THE DECADE, 1911-1921 [1]

District of origin	Upper Chenab Colony	Lower Bari Doab Colony
Jullundhar	7,455	18,799
Sialkot	30,175	8,065
Hoshiarpur	4,401	10,568
Amritsar	12,124	3,099
Gurdaspur	7,512	7,050
Lahore	5,196	13,642
Ferozepur	1,036	4,230
Jhelum	855	1,719
Mianwali	272	37
Hissar	1,228	745
Gujrat	3,107	2,055
Amballa	668	866
Ludhiana	456	1,716
Other districts	1,802	1,450
	76,287	76,141

In Chapters II, III and IV it was shown that the territory now served by the great canals of the Punjab, namely, the Lower Chenab, Lower Jhelum and Lower Bari Doab, was practically waste land prior to the development of the irrigation canal projects. Naturally, these areas were almost devoid of population in the period preceding the British occupation of the province. It was also shown in those chapters that the only lack of the soil in this territory was water, since the structure and content of the soil was favorable to cultiva-

[1] *Punjab Census Report*, 1921, p. 85 *et seq.*
The Upper Chenab Canal Colony serves portions of the districts of Sheikupura and Gujranwala. The Lower Bari Doab Canal serves the Montgomery District.

tion. As early as 1868 the Census Commissioner had stated that " any extension of population can be expected only when the tracts shall have been brought under the ameliorating influences of irrigation and cultivation."[1] To what extent the development of irrigation canals has made possible an increase in the population of the canal-colony areas has been indicated in Table 2 *supra* where it was shown that whereas the average density of population per square mile of the colony areas in 1881 was 148.2, by 1921 it was 243.3 for the colonized areas in general. In Table 6 is shown the remarkable rate of population increase measured in density per square mile for some of the districts in which canal colonies have been developed.

TABLE 6. DENSITY OF POPULATION PER SQUARE MILE IN SIX DISTRICTS WHICH HAVE BENEFITTED LARGELY FROM DEVELOPMENT OF CANAL IRRIGATION PROJECTS [2]

District	1881	1891	1901	1911	1921	Percentage of increase 1881–1921
Montgomery	78	94	97	108	154	98.7%
Shahpur	86	107	109	144	161	87.2%
Mianwali	49	63	56	63	66	34.7%
Lyallpur	12	9	181	261	301	2408.3%
Jhang	113	117	123	152	165	46.0%
Multan	94	107	120	137	150	59.6%

The increasing density of population in the Lyallpur and Jhang Districts is due to the phenomenal development of the Lower Chenab Canal Colony. In the Montgomery District, the increasing population is chiefly due to the completion of the Lower Bari Doab Canal. This district is experiencing a

[1] *Punjab Census Report*, 1868, *op. cit.*, paragraphs 76, 77 and 89.

[2] These figures taken from the *Punjab Census Reports for* 1881, 1891, 1901, 1911 and 1921.

new population increase because of the development of the Nili Bar Colony at the tail of the Lower Bari Doab Canal to which reference has been made in a previous chapter. The increasing density of population in the Shahpur District is to be accounted for by the completion of the Lower Jhelum Canal. The population increase in the Mianwali District is taking place in anticipation of the benefits to be derived from the development of the Thal Canal Project which is now proposed and the completion of the Sind Sagar Railway construction which has brought efficient transportation and communication to this previously isolated region. It is probably true that immigration to the newly irrigated tracts has been chief among the causes leading to the increased density of population in these previously arid regions. Other factors not easily isolated but probably bearing upon the question are the development of better means of communication which lead to the development of new towns in the colony areas, the probable increased fecundity of the indigenous population which tends to become more prolific upon settling down to a more sedentary life such as agriculture makes possible [1] and the possible increase in the size of the colony peasant's family due to the relatively more attractive economic prospects and the more sanitary conditions of the colony tract villages which make possible the survival of a larger percentage of the babies born than in the previous mode of life in the older unsanitary villages from which the colonists were drawn.

That the colony areas attracted large numbers of people other than the peasant colonists is evidenced by the rapid growth of towns and cities in these newly irrigated tracts. Table 7 shows a selected group of colony-area towns with the population of each as the Census Reports of the years indicated present them.

[1] *Lower Chenab Colony Settlement Report* (Lahore, 1904) *passim.*

TABLE 7. GROWTH OF TOWNS IN THE CANAL COLONY AREAS ARRANGED IN ORDER OF SIZE ACCORDING TO 1921 CENSUS RETURNS [1]

Town	1881	1891	1901	1911	1921
Gujranwala	23,661	27,678	30,092	30,307	37,887
Jhang	21,629	33,290	24,382	25,914	30,139
Lyallpur	..	..	9,171	19,578	28,136
Gujrat	18,396	18,050	19,410	19,090	21,974
Sarghoda	..	..	..	8,849	17,728
Chiniot	10,731	13,470	15,685	14,085	17,513
Bhera	15,165	17,428	18,680	15,202	17,027
Montgomery	3,178	5,159	6,602	8,129	14,601
Jalalpur Jattan	12,839	11,065	10,640	11,615	10,792
Khushab	8,989	9,832	11,403	10,159	10,009
Kamalia	7,594	7,490	6,976	8,237	8,916
Hafizabad	..	..	..	..	8,854
Gojra	..	..	2,278	5,417	7,622
Pakpattan	5,993	6,329	5,880	6,334	6,730

The fluctuating population of these colony towns indicates a large degree of mobility of the people in these areas. The fact that towns like Hafizabad do not appear in the census reports at all until 1921 shows that the development of the canal colonies draws trades-people from some of the more settled sections in fairly large numbers within a brief period of time to the new centres where the chances for gain on the part of the middleman are more abundant. In appendix F is given a list of colony towns in which banks have been established for the purpose of facilitating trade and sharing with the people of the colony tracts the relatively better economic life which is there made possible by a more thriving agriculture and a more progressive spirit of economic adventure.

Whether much weight can be given to the suggestion above that the colony peasant may look upon a larger family as an economic asset is somewhat dubious. The writer has not

[1] Figures taken from *Punjab Census Reports*, 1881, 1891, 1901, 1911 and 1921.

found the *Punjabi* peasants of his acquaintance inclined to be particularly foresighted in their consideration of the most economic size of their family. He agrees with the statement made by the Census Commissioner of 1921 that the natural increase of population tends to be accelerated in the newly colonized regions [1] but chiefly on account of the decrease in mortality resulting from better sanitary conditions which the irrigation engineers have practically forced upon the colonists by the care which they have exercised regarding sanitary conditions in the planning of the new villages and towns. The lack of complete data concerning the population problem in the Punjab colonies precludes the isolation of the various factors affecting population growth, in order to give each factor a precise value for purposes of forecast.

This chapter tends to support the conclusions that irrigation canals constructed by the British since 1849 have considerably increased the area under cultivation; that, while population has increased from 16,250,000 in 1868 to 20,685,024 in 1921, the cultivated area of the province has been increasing at an even more rapid rate; the canal-irrigated area had increased from approximately one and a half million acres to over eleven million acres in 1926-27. Thus, the cultivated area per capita has been increased from 1.25 acres in 1868 to 1.42 acres per capita in 1926-27.[2] The proportion of the total cultivated area served by canal irrigation in 1868 was only 6.3 per cent while the percentage of irrigated area to cultivated area in 1921 was 36 per cent. It is probable that the irrigated area served by Government canals in 1926-27 was about 37 per cent of the area cultivated. Thus it would appear reasonable to conclude that not only has the cultivated area per capita expanded but the irrigated area per capita has

[1] *Punjab Census Report*, 1921, p. 47 *et seq.*
[2] On the basis of the 1921 population figures, however.

likewise expanded, making the people of the Punjab far less subject to famine and drought than they were prior to the coming of the British. The statistical materials available for a study of the density of population of the various districts do not give much ground for optimism as to the probable success of the Government in lessening population pressure in the Punjab as a whole by merely shifting the inhabitants from district to district to develop newly irrigated areas. Table 2 shows that there has been only a temporary solution of the population problem. As the gaps caused by emigration from the thickly settled areas are rapidly filled through the natural increase in the population, it is evident that, not only the Government, but the people themselves must soon face the problem of population control. Unless the standard of living in the Punjab is raised and a limit is set upon the size of the family, it is probable that the Government's expansion of the agricultural base can scarcely be expected to continue to outstrip the population increase. Our study of this situation supports the conclusion that the Government is probably engaged in a losing battle, since the population will probably increase more rapidly than will the available means of subsistence. If this proves to be the case it will require but a few generations for the population of the Punjab to press as heavily upon the available food supply as it did prior to the inauguration of the British irrigation schemes in the province. This conclusion is almost unavoidable in view of the fact that, aside from agriculture, there will probably never be any large development of industry in which any considerable portion of the population will be able to earn a livelihood. Brilliant, therefore, as the Government's achievements may appear on the technical and financial side, the resulting *economic* benefits must remain doubtful unless we hold the view once expressed by Professor Pigou that " an increase in numbers, even though it leave economic welfare per head the same, involves an addition to economic welfare in the aggregate."[1] Considering the problem in its long period aspects, the temporary gains in physical well being may well prove to be but a means of increasing the misery of larger numbers in the future. Without some control of the growth of population in the Punjab the advantages of irrigation to increase the amount of food produced will prove entirely inadequate. A study of the population figures for the Punjab since 1868 presents small basis for complacency in regard to this aspect of the problem.

[1] Pigou, A. C., *Wealth and Welfare* (London, 1912), p. 29.

CHAPTER VI

THE EFFECT OF IRRIGATION ON THE WEALTH AND AGRICULTURE OF THE PUNJAB

AGRICULTURE is the predominant industry of the people of the Punjab. As in the rest of India, the village is the typical agricultural unit in this province also. The rural population of the Punjab is scattered in 34,000 villages whose average size is just under 500 inhabitants. The agricultural holdings of the peasants are small in the Punjab as a whole and except in certain cases in the canal colonies do not furnish full-time employment to the peasant population. These many villages are built very much on the same pattern, or lack of pattern, and a visit to the villages of the province soon impresses the traveler with a sense of monotony. Mud huts with low flat roofs predominate. Occasionally one sees a cottage built of bricks signifying the habitat of one of the economically more fortunate peasants. In many villages the houses of the money-lenders and shop keepers are decidedly better built than those of the average peasant villager. The huts generally form part of a wall which encloses a small space or courtyard which houses the peasant family's moveable property. Not only will the enclosure prove the habitat of the family but also of the goats, chickens, a water buffalo or two, several oxen and the ever present nondescript dog. It is most unusual to find a hut with windows and even less often does a window boast of glass panes. Most of the houses are built as one room into which there is but one opening, the door, which must serve as the home's only means of ventilation. During the cold season, from the middle of

November to the end of April, the door to the house is kept closed as tightly as possible. During this period when there is the minimum of work to be done out of doors the family spends most of its time in the poorly ventilated atmosphere of the one-room house, hovering in the center of the room about the cowdung fire which serves the double purpose of furnace and cooking stove. The peasant's hut is almost entirely free of furniture. Chairs are unnecessary and tables quite unknown. The *charpai* [1] in one or more corners of the room provides a place to sleep as well as an inviting lounging place during the long winter season. The stove, if such it can be called, consists of a rudely constructed pit with a hardened clay wall. Sometimes the pit is covered with a bit of tin or iron, but generally it is open and the cooking utensils of the housewife are placed upon the smouldering cowdung fire. Those living thus cooped up in a stifling atmosphere made still less bearable by the incessant smoke from the fire, for which there is no chimney, must find such homes fruitful sources of eye irritation leading often to serious diseases of the eye. It is probable that this torture of the eyes in the huts during the winter coupled with the excessive glare of the summer's sun cause the large amount of eye infection which is one of the superficial features noted by casual travelers through the Punjab.

The inner group of village houses is generally occupied by the peasants who till the village land. The outer houses of the village are occupied by the village menials and artisans. The outer houses or huts also accommodate any additional bullocks which may not find shelter in the enclosures of their peasant owners. However, as far as possible, the oxen and cattle of each peasant tend to be kept as close to the owner's habitation as space will permit. This proximity of

[1] The *charpai* is the rudely constructed four-cornered native bed made of a wooden frame covered with a coarse fibre-rope woven net.

cattle to the hut, or within the peasant's small crowded courtyard, makes for the economical collection and utilization by the household of every bit of barnyard manure. Each peasant hoards the barnyard manure which he is able to collect since this forms the chief article of fuel for cooking purposes. As has been indicated in a previous chapter, the failure on the part of the peasant to utilize this manure in renewing the fertility of the soil is probably one of the chief causes of India's poverty.[1] It will be remembered from a reference to the proposed development of plantations along the irrigation canal banks, that the British have long been attempting to increase the fuel supply of the province in order to permit the use of manure on the fields. Thus far, however, the peasants have not given up their preference for use of cowdung as fuel.

The Punjab village is not an inviting place. It is generally quite bereft of order, its streets wandering crookedly about in order to miss the mud huts and walls of the various individual compounds of which the village is composed. The manure piles, carefully husbanded for future fuel use, and the dung-plastered walls of the huts give off an obnoxious odour and attract multitudes of flies, especially in the hot season. Fortunate is the village which boasts of a tree or two. Where such exist, they become the favored meeting place of the old men of the village who will bring out their *charpais* and spend the long summer days smoking the *hookah*[2] while they discuss the petty gossip and social prob-

---

[1] The view is generally held that it is the absence of a sufficient supply of firewood which, over large parts of India, compels the burning of cowdung as fuel. But it must be recognized that there is often a definite preference for this form of fuel, as its slow-burning character is regarded as making it especially suitable for the needs of the Indian housewife.... Our evidence does not suggest any alternative fuel for domestic purposes in districts where wood and coal are dear.—*Royal Commission on Agriculture in India, op. cit.,* p. 83.

[2] The *hookah* is the Indian pipe used by the men. It is a quite cleverly

lems of the village in the shade of the village tree. Water is a scarce article and hence the village is bereft of vegetation of greenery of any kind which might relieve the cheerless aspect of the village environment. During five years' residence in the province the writer failed to see a single green grass plot kept by a native Indian for his family in a Punjab village. While the old men of the village find solace in meeting incessantly either under the one or two shade trees of the village or in the shade of some fortunately placed high mud wall, the village women find plenty of occasion to meet and gossip at the village well from which the water for all purposes must be drawn. The division of labor has long since dictated that the task of drawing water at the village well shall be predominantly the work of women. Fortunately for the villagers of the Punjab, the summer season makes out-door life far more endurable than life indoors could possibly be, with the result that comparatively little time is spent during the hot months in the stuffy huts. Were this not the case the mortality rate among the population of the villages would undoubtedly be higher than it is. For the weak and the sick the village presents a difficult environment. Thus, in a recent study of a particular village in the Punjab the investigators wrote:

> Every possible circumstance seems to be present to hinder recovery or to shorten life. If the patient is suffering from wounds, the chances of septic poisoning are very great. For a woman recovering from childbirth, the chances of her getting puerperal fever must be much greater even than they are in

constructed utensil which permits the 'straining' of the tobacco smoke through water prior to inhalation by the smoker. When smoked by a veteran, the gurgling sound of the bubbling water indicates that the smoker is well content. A familiar sight in any Punjab village of a hot summer's day is the contented puffing of the elders of the village at ease with their *hookahs*.

Indian cities. . . . In the circumstances, it is not to be wondered at that the mortality among women is higher than it is among men, or that the mortality among children under one year can only be described as appalling. The desire of the young married pair is to have a male child. A child is born, and even if it is a male, its chances of surviving the first year are only three to one; it dies and the woman must suffer the pangs of maternity a second and a third and a fourth time. . . . The unsanitary conditions necessitate a heavy birth-rate, and a heavy birth-rate, owing to the risks run at each birth, causes a heavy death rate among women.[1]

In the Punjab, the village site is generally located somewhere near the centre of the land tilled by the villagers. The land may be owned by some wealthy landlord and thus form a portion of a large estate. The land surrounding the village may be owned by members of a joint family, giving rise to a form of ownership sometimes designated as collective. In a province in which the Hindu joint family system has not yet been entirely eliminated such a " joint holding " may belong to a large number of individuals, all members of one joint family. This perhaps has given rise to the assumption on the part of certain writers that land is communally owned in contrast to private ownership. Occasionally writers have insisted that each village has " common lands " which are the joint property of the members of the village community as a whole. The writer has not been able to discover in the Punjab any definite case of village lands owned in common by all the villagers. What he has found are certain lands set aside as " village commons " but these common lands are available for use only to the owners of specific portions of the remainder of the land which the village tills and the rights to use the village common land are parcelled out in proportion to the amount of land owned in the cultivated

[1] Singh, Gian and King, C. M., *Ghaggar Bhana* (Lahore, 1928), pp. 2-3.

tract. This is evidently an entirely different form of property right from that of purely collective land ownership. Most common in the Punjab are the village tracts which are owned individually by the various individuals comprising the village community. These holdings of the private owners of cultivated land have become quite small in the older sections of the province where the Hindu law of inheritance has had ample time to cause minute fragmentation of holdings among the surviving sons of the deceased landowner. It has been shown in previous chapters that the development of irrigation and the launching of colonization schemes has tended to create an original allotment of some twenty-two and a half acres or more per colonist at the time of the colony's launching. Thus the holding in the canal colonies might be expected to be fairly large in comparison with the average size of the holding in the province as a whole. Also, the larger size of the holdings in the colony tracts would be expected to increase the average size of the Punjab holding. The average size of the holding in the Punjab in 1926-27 was 6.7 acres per owner. The average size of the holding in the canal colony districts was 9.8 acres per owner. Table 8 presents the average holding per cultivator in the various districts of the Punjab.

The table indicates the relative intensity of population pressure upon the soil as measured in average area cultivated per owner or tenant. The most significant feature of the table is the amazing shrinkage in the size of the holding even in the canal colony districts where a few years ago (in no case prior to 1882) the size of the holding was not less than 22½ acres on the average but in 1926-7 the average holding had been reduced to 9.8 acres. If this shrinkage continues at the rate prevailing since colonization began, the advantage of the colony tracts in so far as average size of holding per cultivator is concerned will soon be entirely elimi-

CANAL IRRIGATION IN THE PUNJAB

TABLE 8. AVERAGE SIZE OF CULTIVATORS' HOLDINGS IN THE PUNJAB[1]

Districts	Area per cultivator Acres	Districts	Area per cultivator Acres
Hissar	11.3	Rawalpindi	2.8
Rohtak	5.7	Attock	8.8
Gurgaon	4.8	Mianwali	9.6
Karnal	6.5	Muzaffargarh	5.3
Ambala	4.2	Dera Ghazi Khan	5.5
Simla	1.1	Lahore	8.7
Kangra	1.8	Gujranwala	7.2
Hoshiarpur	2.0	Sheikhupura	10.6
Jullundhar	5.2	Gujrat	5.0
Ludhiana	6.0	Shahpur	11.6
Ferozepore	10.5	Montgomery	9.1
Amritsar	5.3	Lyallpur	14.8
Gurdaspur	3.2	Jhang	10.7
Jhelum	5.1	Multan	10.0
Average for the Punjab	6.7		
Colony district average	9.8		

nated. This would tend to support the conclusion stated at the close of Chapter V that the development of irrigation schemes making possible the cultivation of otherwise arid waste lands is not a permanent solution of the problem of population pressure on the means of subsistence. At the same time it needs to be emphasized anew that the people of the Punjab are probably better fed today than they were prior to the coming of the British due to the relatively more certain and larger production per acre made possible by the increase in the irrigated area in relation to the total cultivated area.

[1] Data for this table were compiled from the various district gazetteers for 1926-27. In the Punjab very few tenants are without any land of their own, hence the average area per cultivator tends to be practically identical with the average size of the holding per land owner. A similar conclusion has been reached by those investigators who worked out the *Size and distribution of agricultural holdings in the Punjab* under the auspices and direction of the Board of Economic Inquiry of the Punjab. See p. 4 of their report issued in 1925.

IRRIGATION AND WEALTH 103

Prior to the annexation of the Punjab by the British, private property in land was not a very valuable asset due to the unsettled political conditions of the times, the absence of a stable government capable of establishing peace and protecting property rights, the revenue demands of the Government, which fluctuated widely the absence of civil courts and a code of laws on the basis of which property rights could be adequately defended. One of the first tasks of the British in the Punjab was to record, define and legalize certain forms of land tenure which they found in existence in the province with a view to levying the land tax which promised to be the most fruitful source of government income. The Government realized that in order to secure the revenue on which the State was to subsist it must also assure private property rights in land to the peasants, if wealth and prosperity were to be increased.

The British Government, while determining to limit and render moderate its own demands on the land, and to give up forever the inordinate pretensions of the rulers it superseded, found itself face to face with the task of giving legal security and definition to various degrees and kinds of right or interest in land.[1]

It is not surprising to discover that with the establishment of a stable government and the systematic recording of titles to land, land began to achieve a trading value and frequent changes in land ownership took place. Land values tended to rise. Mr. M. L. Darling has emphasized at great length in his book [2] the point that the Punjab peasant made the interesting discovery that land which he owned provided a basis for credit and that with the increasing utilization of land as

[1] Baden-Powell, B. H., *Systems of Land Tenure in British India* (London, 1896), p. 65.
[2] Darling, M. L., *op. cit., passim*.

security for loans, the peasant naturally fell into the clutches of the money lender. This led to a rapid rise in the importance of the money lender in the province and loans which could not be repaid were liquidated by title to the land passing to the money lender. This concentration of land ownership in the hands of the money lenders caused the British to pass the Land Alienation Act of 1901 for the Punjab.[1]

The extension of irrigation has undoubtedly added still further to the value of land as the basis for loans and for outright sale. The Punjab Land Alienation Act has limited the extent to which land may pass into the possession of the money lenders but has tended to make it practically impossible for a non-agriculturist to gain control of agricultural land. The Government was insisting upon the necessity of limiting the allotment and sale of land in the colony areas to members of the agricultural classes in order to insure the most likely success of the colonization schemes. The passing of such a law shows the Government's interest in maintaining the Punjab as a province of small landed proprietors rather than permitting the ownership of land to become centralized. No very complete study of the changing value of land in the Punjab has been made. However, on the basis of information contained in the annual *Land Revenue Reports* of the Punjab Government, Mr. Calvert has presented the material embodied in Table 9:

---

[1] *Royal Commission on Agriculture in India, op. cit.*, pp. 433-4. This law made the sale of agricultural land in execution of a decree illegal; it stated that without special governmental sanction agriculturists could not sell their land to non-agriculturists; it declared that the conditional sale clause in land mortgages were illegal; it required that mortgages of land by agriculturists in favor of non-agriculturists provide for automatic redemption; and finally, in order to prevent evasion of its provisions, declared that land may not be leased to non-agriculturists for a period longer than five years. *Cf.* Calvert, *op. cit.*, p. 136.

TABLE 9. AVERAGE PRICE PER ACRE OF CULTIVATED AGRICULTURAL LAND IN THE PUNJAB [1]

Year	Price
1869–70	Rs. 10
1875–76	20
1880–81	18
1885–86	30
1890–91	61
1895–96	59
1900–01	77
1905–06	85
1910–11	124
1916–17	227
1919–20	275

Table 9 indicates a fairly constant rise in the value of cultivated land which entered the market. The question as to the basis for this increased value of land is one not easily answered. As shown in an earlier section, the cultivated area of the Punjab has tended to increase at a slightly more rapid rate than population has increased. Thus the cause of the mounting price of cultivated land is not to be found in the increased pressure of population upon each cultivated acre. Actually, between 1868 and 1921 the cultivated area per capita has expanded from 1.25 acres to 1.41 acres. Nor, as will be shown in a later section of this chapter, is the increase

[1] Calvert, H., op. cit., p. 101. See also table in Appendix B showing average price of colony-irrigated land sold at auction.

The method used in determining the average price of cultivated land has been to take the land which is sold each year, ascertain the price paid per individual parcel, then sum up the total paid and divide by the number of acres sold to get the average price paid per acre. Following that method the Revenue Department in 1927 showed that the average price per acre of cultivated land sold during that year was Rs. 438. It is not at all certain that this method indicates the average value of the cultivated land of the Punjab since it takes into consideration only that portion of the land which enters the market during the year in question. The land which is sold each year may or may not be a fair sample of the agricultural land under cultivation in the Punjab. But that is the only method thus far employed in ascertaining the average price per acre.

in land values due to any large extent to the adoption of better methods of agriculture. The gross produce of the soil has probably increased considerably because a larger proportion of the cultivated area is now under irrigation. But this increased production per acre due to irrigation has not resulted in a fall of prices of agricultural products or in a fall in the price of land. Mention has been made of the limited size of the average holding in the Punjab. The average size of the cultivated holding has been given as 6.7 acres for the whole province. This however does not mean that the tenant is able to cultivate a plot as large as 6.7 acres. Because most of these holdings are fragmented in successive generations in accordance with the operation of the Hindu law of inheritance, these holdings are greatly reduced in size. Even though the total holdings cultivated by tenant or owner may be equal to 6.7 acres on the average, many of the plots cultivated are less than 1/20 of an acre in size. In addition to this hindrance due to size, the fragments lie within the village plot of land and the many small fragments and individual plots are not enclosed. Hence, the total village land area must be cultivated according to custom and not in accordance with the abilities and desires of the individual owners. It is still necessary in the Punjab village to consider the inconvenience of crossing the many fields of one's neighbors in order to gain access to the particular small plot to be worked upon on a given day. But, while the plots are not enclosed they are carefully marked by boundary limits in the form of stakes or blocks of stone. Thus, while the plots are necessarily cultivated according to the custom of the village peasants, the village land is not worked as a unit. Each individual peasant tends his own plots. He necessarily wastes much time and energy going from one of his scattered plots to another. The obvious solution to the problem of excessive fragmentation is consolidation of holdings. But in the successful carrying to fruition of a plan to consolidate holdings among these Indian peasants the would-be-reformer meets with all the inertia founded on long traditional practice. Land ownership as such, regardless of its profitable or uneconomic nature, is a matter of social prestige in the Punjab. The desire on the part of the people to own land whether in economic units or not has been one of the strongest factors in forcing the price of land higher in each successive decade. Consolidation of holdings is one of the important goals of the cooperative movement among the peasants of the Punjab sponsored and supported by the British Government.[1]

It is possible that, owing to fragmentation, the rise in land values in the Punjab has been somewhat less than it might otherwise have been; this practice of subdivision has, on the other hand, tended to increase the demand for land. Since many of the holdings have been reduced to a size that is economically unprofitable, the existence of so many potential nuclei for economic holdings produces a continuous demand for land on the part of the small holder. The number of fragments places the purchase of such units within the reach

The total area consolidated was over 64,000 acres, nearly twice as much as in the preceding year; but the rate of progress is still too slow. . . . If the increase in value resulting from consolidation is reckoned at the low figure of Rs. 50 per acre the total increase in capital value of land consolidated during the year would amount to 32 lakhs of rupees. If consolidation became universal in the province the benefit to the cultivators would be incalculable.—*Cooperative Societies Report In the Punjab For Year Ending July 31, 1928*, p. 5.

As the result of patient work which has now extended over eight years, the movement for consolidation in the Punjab has assumed the dimensions of an important agricultural reform. . . . The total area dealt with in the first five years was 39,757 acres; in the following year alone the area consolidated was over 20,000 acres, and last year (1926-27) the area was over 38,000 acres.—*Royal Commission On Agriculture Report, op. cit.*, p. 139.

of small holders, whose bidding, however, tends to force the price of land upward. But this statement does not invalidate the statement just made that consolidation of holdings now tends to increase the price of land. The point which we wish to stress is that both fragmentation and consolidation, since they occur in the province at the same time, are forces affecting the price of land. But the action of fragmentation upon the price of land can be understood only by keeping in mind the Punjab peasants' attitude toward land ownership as a sign of social prestige. It is obviously impossible to measure, in any statistical manner, the effect of this attitude upon price but its influence cannot be ignored.

Thus far the problem of the increasing value of cultivated land has been considered from the point of view of the province as a whole without regard to the differences in output per acre on irrigated and non-irrigated land. The increasing areas brought under irrigation must have had considerable effect upon the rising average price of cultivated land in the Punjab. In Tables 10, 11 and 12 data relating to the average produce per acre of irrigated and non-irrigated areas in the various districts of the province are given for the more important commercial crops, wheat, cotton and sugar cane.

Tables 10, 11 and 12 indicate that the irrigated area under cultivation in relation to these three crops, wheat, cotton and sugar cane has an advantage over the unirrigated areas under cultivation ranging from approximately thirty to one hundred per cent measured in produce per acre. If the lower percentage be taken as a conservative measure of the advantage enjoyed by the producer utilizing irrigated land it would appear that the province is both absolutely and relatively better situated in relation to food supply now than it was prior to the development of the vast irrigation schemes by the

IRRIGATION AND WEALTH

TABLE 10. AVERAGE YIELD OF WHEAT IN POUNDS PER ACRE ON IRRIGATED AND NON-IRRIGATED LAND IN SOME OF THE IMPORTANT WHEAT PRODUCING DISTRICTS OF THE PUNJAB[1]

District	1901–02	1906–07	1912–13	1917–18	1923–24
Lahore					
Irrigated	752	1080	560	960	1000
Unirrigated	382	720	520	600	520
Gujranwala					
Irrigated	..	840	820	960	960
Unirrigated	..	560	460	560	560
Gujrat					
Irrigated	800	800	900	900	1000
Unirrigated	560	560	600	600	800
Shahpur					
Irrigated	960	960	840	960	930
Unirrigated	800	800	600	640	650
Montgomery					
Irrigated	800	800	760	960	960
Unirrigated	600	600	600	600	600
Lyallpur					
Irrigated	..	960	1000	1040	1200
Unirrigated	..	480	480	480	500
Jhang					
Irrigated	840	840	820	900	960
Unirrigated	560	560	600	600	600
Multan					
Irrigated	960	960	960	900	960
Unirrigated	720	720	750	600	600

[1] Figures taken from the district gazetteers for the years specified. The figures show a tendency toward a stability of output per acre for each separate district over the years indicated. This would tend to create a suspicion as to care taken in making the figures representative of the facts of each year's crops. However, the whole series of data shows a fairly constant ratio of production on irrigated land in comparison with unirrigated land. The decidedly favorable production statistics relating to irrigated land leave little doubt as to the beneficial effects of irrigation on the land. This undoubtedly is reflected in the selling price of irrigated land. See Appendix B giving prices of canal colony tracts sold at auction over a period of years.

TABLE 11. AVERAGE YIELD OF COTTON IN POUNDS PER ACRE ON IRRI-
GATED AND NON-IRRIGATED LAND IN THE MORE IMPORTANT
COTTON-PRODUCING DISTRICTS OF THE PUNJAB [1]

District	1906–07	1912–13	1918–19	1924–25
Lahore				
Irrigated	440	440	440	440
Unirrigated	240	240	240	240
Gujranwala				
Irrigated	400	340	340	467
Unirrigated	240	253	267	365
Gujrat				
Irrigated	480	480	480	480
Unirrigated	300	300	333	333
Shahpur				
Irrigated	520	440	483	467
Unirrigated	260	220	247	247
Montgomery				
Irrigated	400	380	533	600
Unirrigated	240	230	233	240
Jhang				
Irrigated	400	357	443	333
Unirrigated	240	257	257	200
Multan				
Irrigated	320	320	333	333
Unirrigated	220	220	220	220

[1] Figures taken from the district gazetteers for the years specified. For the Lyallpur district no figures are available for average outturn per acre on unirrigated land, but for irrigated land the average outturn of cotton per acre for the four specified years in the table was 360, 413, 583 and 533 pounds respectively. The steady increase in average yield per acre in the Lyallpur district is to be accounted for largely by the fact that the district is one of the most favorably situated in relation to irrigation and by the intensive experimentation for increasing the efficiency of agriculture which is carried on at the agricultural college located in this district. Furthermore, the preponderant majority of the inhabitants of this district settled there as colonists, thus providing a quite pliable group of agriculturalists more ready to try out the successful methods of agriculture which the experts at the Government Agricultural College had worked out experimentally. More is said about the Government's program of agricultural experimentation at the close of this chapter.

TABLE 12. AVERAGE YIELD OF SUGARCANE IN POUNDS PER ACRE ON IRRIGATED AND NON-IRRIGATED LAND IN THE MORE IMPORTANT SUGARCANE PRODUCING DISTRICTS OF THE PUNJAB [1]

District	1901-02	1907-08	1912-13	1916-17	1924-25
Lahore					
Irrigated	1700	1700	1700	1700	1700
Unirrigated	1000	1000	1000	1000	1000
Gujranwala					
Irrigated	1170	1170	1280	1280	1600
Unirrigated	800	800	1070	1070	1070
Sheikhupura					
Irrigated	..	..	..	..	1400
Unirrigated	..	..	..	..	1070
Gujrat					
Irrigated	1040	1040	750	1300	1300
Unirrigated	788	780	650	780	780
Shahpur					
Irrigated	1600	1600	850	1600	1600
Unirrigated	400	400	..	400	400

British. As has been shown in previous chapters, the cultivated area per capita in 1868 was but 1.25 acres. Of this cultivated area only 6.3 per cent was served by canal irrigation. By 1921 the cultivated area per capita had been increased to 1.41 acres while the proportion of irrigated to non-irrigated land had increased, no less than 36 per cent of the total area under cultivation then being served by canal irrigation. Thus it may be concluded that not only was there a larger area per capita available for the production of

[1] Figures taken from the district gazetteers for the years specified in the table. The Lyallpur district gazetteer gives no average outturn per acre for the unirrigated area. For the irrigated area the average yield of sugar cane per acre varies from 1500 to 1800 pounds depending upon the general condition of the crop due to vagaries of the season other than water supply, which in this district is fairly stable since the extensive development of perennial irrigation canals throughout the district.

agricultural produce, but the average production per acre had been increased by about 30 per cent on account of the more bountiful production made possible by the irrigation service. The data presented in Tables 10, 11 and 12 tend to support the conclusion of the Royal Commission on Agriculture in India regarding the possibility of soil deterioration in the Punjab.[1] In most cases as indicated by the tables cited, there has been a slight tendency toward an increased production of produce per acre in the Punjab on the irrigated areas while the unirrigated areas show a tendency to increase also. This increased production on the unirrigated areas is probably due to the fact that new lands contiguous to the newly irrigated tracts have been brought under irrigation and have reaped an indirect and statistically unmeasurable benefit from the irrigation canals as was suggested in Chapter III. But the increased production made possible by the extension of irrigation and the development of transportation facilities was not sufficiently large to account for the marked increase in the price of land which has been discussed in the preceding section of this chapter.

An examination of the value of crops raised on the irrigated land of the Punjab will next be made with the purpose of discovering the relationship between produce per acre and the price of land. Table 13 gives details which show the total area assessed to water rates and the total value of the crops grown in 1926-27 estimated at the prevailing prices of that year.

The total estimated value of crops raised on the government canal irrigated areas in the Punjab in 1926-27 was Rs.

[1] "We have been informed that, in Bombay, Bengal and Burma there is no evidence of any decline in the yield of staple crops, while the local governments of Madras, the United Provinces and the Punjab tell us that the tendency in those provinces is towards a slight increase in outturn." *Royal Commission on Agriculture Report, op. cit.*, p. 75.

45,77,53,509.[1] The area on which water rates were assessed was 10,073,728 acres. The average value of the crops on the irrigated acre in the Punjab was therefore approximately Rs.

TABLE 13. AREAS IN ACRES ASSESSED TO WATER RATES FOR THE IMPORTANT COMMERCIAL CROPS OF THE PUNJAB WITH THEIR TOTAL MARKET VALUE IN TERMS OF PRICES PREVAILING DURING 1926-27

Name of crop	Area assessed to water rates	Total value
Cereals:		
Wheat	3,377,431 acres	Rs. 20,43,87,917
Barley	103,490 "	36,83,100
Rice	404,534 "	2,45,81,790
Maize	242,963 "	1,18,33,481
Oats	33,134 "	12,36,040
Mixed grain	253,184 "	81,91,850
Millets & pulses		
Great millet	288,028 "	92,43,825
Spiked millet	286,994 "	72,89,990
Gram	471,117 "	1,56,98,458
Fibres		
Cotton *desi*	961,438 "	2,86,13,498
Cotton American	721,799 "	2,33,51,115
Oil seeds		
Gingelly or sesame	20,307 "	6,84,786
Rape (*sarson*)	95,432 "	42,85,922
Rape (*toria*)	461,944 "	2,04,68,194
Linseed flax (*alsi*)	1,245 "	55,755
Linseed flax (*taramira*)	21,638 "	7,71,797
Turnips	231,100 "	88,47,935
Lucerne grass (*senji*)	390,575 "	1,05,18,507
Sugarcane (*ganna*)	205,474 "	2,82,80,949
Sugarcane (*shaftal*)	18,900 "	9,45,000

[1] *Irrigation Report*, 1926-27, *op. cit.*, part ii, p. 25.

[2] Figures taken from *Administration Report, Irrigation Department, Punjab Government* (Lahore, 1927), pp. 24-25. The irrigation report lists other minor crops such as tobacco, poppy, tumeric, aniseed, camin, rosella hemp, false hemp, corlander, indigo, henna, safflower, fruits and vegetables. These crops while important in their aggregate value are not produced on a large scale. Many of them are produced for local consumption rather than for the market.

45. If this agricultural produce had been equally distributed among the people of the province there would have been available for distribution approximately Rs. 20 per head. Since 36 per cent of the cultivated area is served by government irrigation canals and since production on the irrigated land is conservatively estimated as approximately 33 per cent more per acre than on non-irrigated land, it would seem reasonable to conclude that approximately one-half of the total produce of the Punjab in 1926-27 was raised on the irrigated lands. If this be conceded, the per capita income of the population of the province for 1926-27 measured in agricultural produce at its market value for that year would amount to less than Rs. 45 or about $16.00. If the agricultural population alone is considered, approximately ninety crores [1] of rupees, if equitably distributed among the 85 per cent of the total population of the province thus engaged, would made the per capita share Rs. 50 or $18.00. This indicates that the standard of living in the Punjab is still low and that in spite of the increased development of irrigation and some slight advance in agricultural efficiency the production of wealth is not sufficiently large to permit a very great rise in the standard of living of the people. This fact may be shown by means of a table indicating average yields of the important crops grown in the province. Table 14 presents data of that type for eleven agricultural crops on the basis of estimates collected and prepared by the Punjab Government. Table 14 indicates that there is a considerable degree of variation in the normal yield per acre of the various crops shown as estimated by the Government. The irrigation department officials have worked out a normal yield per acre for each of the important canal areas. The basis of such an estimate is an average production over a period of years characterized

---

[1] A *crore* is equal to 10,000,000. Thus 90 crores would be 900 million.

by what may be called normal rainfall and normal weather conditions. Normal yields for barley tend to vary but little in the various canal areas, the highest yield among the esti-

TABLE 14. ESTIMATED NORMAL OUTTURN PER ACRE IN MAUNDS OF ELEVEN IMPORTANT CROPS PRODUCED IN THE PUNJAB CANAL AREAS [1]

Canal area	Wheat	Barley	Gram	Great millet	Cotton *desi*	Cotton American	Sugarcane	Rice	Maize	Spiked millet	Kabi oil-seeds
Western Jumna	13.6	13.6	9.9	6.8	6.6	5.8	32.1	11.5	13.7	5.8	6.6
Sirhind	12.0	13.0	11.0	5.9	5.0	4.5	24.4	11.5	16.3	5.0	6.1
Upper Bari Doab	12.0	12.2	10.5	5.9	5.8	5.4	24.6	21.8	15.4	5.0	6.4
Lower Chenab	14.5	12.2	7.6	7.6	6.3	5.8	20.9	19.7	14.2	8.3	8.0
Lower Jhelum	11.5	13.4	8.0	9.0	5.6	5.5	19.5	19.7	12.2	8.3	7.3
Upper Chenab	15.2	12.2	7.6	7.3	5.3	5.1	18.3	20.1	11.0	8.3	6.4
Lower Bari Doab	11.5	12.2	6.8	6.7	6.9	6.1	20.7	18.3	15.9	8.3	6.1
Upper Sutlej	11.8	12.2	7.7	6.7	7.1	7.2	20.7	18.3	15.9	8.3	6.1
Sidhnai	11.0	12.2	7.7	6.7	4.1	4.1	20.7	18.3	15.9	8.3	6.1
Indus	9.8	12.2	7.7	4.6	4.1	4.1	20.7	9.8	15.9	8.3	6.1
Shahpur	11.5	13.4	8.0	9.0	5.6	5.5	19.5	19.7	12.2	8.3	7.3
Muzaffargarh	8.5	12.2	7.7	6.7	3.3	3.3	20.7	12.2	15.9	8.3	6.1
Upper Jhelum	12.2	13.4	7.3	9.0	5.9	5.8	15.9	19.7	12.2	8.3	5.9
Pakpattan	11.5	12.2	6.8	..	..	..	..	..	..	..	6.1

mated normals which may be expected in the various areas being found in the case of the Western Jumna canal area. But the high normal of 13.6 maunds is only 1.4 maunds above the low normal as worked out for the areas served by the Lower Chenab, Upper Chenab, Lower Bari Doab, Upper Sutlej, Sidhnai, Indus, Muzaffargarh and Pakpattan canals. The low normal is thus 12.2 maunds for barley. In the case of rice the normal yields vary greatly, from a high estimated normal of 21.8 maunds per acre in the Upper Bari Doab region to a low normal of 9.8 maunds in the area served by

[1] *Irrigation Report, op. cit.,* part ii, pp. 26 *et seq.* One maund is the equivalent of 80 pounds in the Punjab.

the Indus inundation canals. In Chapter VII it will be shown that this variation in the normal yields of staple crops in the various canal areas tend to place a very uneven burden upon the peasant since water rates are assessed on the basis of area cultivated for each crop, but the water rate is not made to vary with the yield except in cases where there is practically a total crop failure when the Government remits the water rate for that particular year.

Normal estimated yields of wheat vary considerably from a high normal of 15.2 maunds (19.1 bushels) per acre in the Upper Chenab canal tract to a low of 8.5 maunds (10.43 bushels) in the Indus Inundation canal tracts. The total acreage devoted to wheat culture in the canal-irrigated areas of the Punjab in 1926-27 was 3,377,431 and its estimated value was Rs. 20,43,87,917.[1] Thus out of a total area assessed to water rates for that year, slightly more than one-third was devoted to wheat. Wheat accounted for just about 45 per cent of the value of crops produced in the irrigated portions of the province in 1926-27. Wheat is the predominant commercial crop, not only of the irrigated portions of the Punjab but also in the older unirrigated sections. Thus wheat continues to be raised even in those parts of the Punjab where the land is minutely fragmented. This necessarily reduces the potential yield of wheat in the province since the small fragments of land can be worked only at a disadvantage. Wheat is a crop for extensive cultivation. It requires comparatively little work to produce a fairly good yield and does not respond as vigorously as certain other crops to intensive human effort to increase the yield. Wheat is the staple article of food of the people of the Punjab. But because wheat is not extensively used as food in other provinces of India, the Punjab generally finds small scope for exportation of surplus wheat produced within its bound-

[1] *Irrigation Report, op. cit.*, p. 25.

IRRIGATION AND WEALTH

aries to the other Indian provinces. Therefore, most of the wheat exported from the Punjab must be shipped to Europe via its nearest seaport, Karachi. The figures concerning wheat exports from Karachi are therefore significant in showing the surplus wheat which the Punjab ships abroad. Wheat exports from Karachi in 1925-26 were 154,224 tons; in 1926-27, 206,478 tons and in 1927-28, 318,593 tons.[1] Practically all of the wheat shipped from Karachi enters that port from the Punjab. That this province has wheat to export implies that in spite of the comparatively large consumption of wheat in the province, many of its people must use some of the cheaper grains as their basic food.

A study of the tables in this chapter clearly shows that the production of food products per acre, even in the irrigated areas, is still very low. The experimental farms of the Government have been able to produce extremely high yields of Durum wheat per acre, reaching 39 maunds,[2] which would seem to indicate that the soil, climate and water supply from irrigation canals are quite favorable to a much higher normal yield in the irrigated areas than even the highest normal of 15.2 maunds thus far reached in the Upper Chenab colony region. Low yields are largely due to the prevailing inefficient methods of agriculture employed by the Punjab peasant even in the canal colonies.

Cotton ranks second among the commercial crops of the Punjab as is shown by Table 13. The estimated normal yield of *desi* (native) cotton ranged from a low normal of 3.3 maunds (264 lbs.) to a high normal of 7.1 maunds (568 lbs.) while for the American cotton [3] the normal yields varied

---

[1] *Administration Report, Karachi Port Trust*, 1927-28, p. 4.

[2] *Administration Report Punjab Government*, 1926-27, p. 57.

[3] A long fibre cotton which the Government is attempting to popularize among the Punjab peasants.

from 3.3 maunds (264 lbs.) to 7.2 maunds (576 lbs.) per acre. Government experimental farms have been established in various parts of the Punjab [1] to aid the peasant in developing more modern methods of agriculture. Cotton produced in the canal irrigated portions of the Punjab in 1926-27 was valued at Rs. 5,19,64,613 thus accounting for about 11 per cent of the gross value of the canal-irrigated produce. Sugarcane is the third important commercial crop of the Punjab as Table 13 shows. In its production in the canal-irrigated tracts there is a variation in the normal yields from a low of 15.9 maunds per acre to a high estimated normal yield of 32.1 maunds. The total irrigated area under sugarcane culture in 1926-27 was 224,374 acres and the crop was valued at Rs. 2,92,25,940, thus accounting for approximately 6.5 per cent of the value of produce raised on the canal irrigated tracts.

The Punjab Government has established its Agricultural Department headquarters in Lyallpur, one of the thriving towns in the canal colony area. The Government Agricultural College is also located there. Hence, Lyallpur is the centre for a constant stream of propaganda favoring better agricultural methods and the source of many of the experiments which the Government is inaugurating in the effort to develop a more efficient agriculture. The Government's program may be roughly outlined under the following heads: (i.) to provide a source and supply of pure seeds for the peasants at prices which will not prove prohibitive; (ii.) to introduce by means of practical demonstration on its own agricultural experimental farms new methods of producing better crops, the use of implements especially adapted to the needs of the Punjab peasant and (iii.) to make an incessant educational effort to keep the achievements and suggestions of the agri-

---

[1] See list of experimental farms which the Agricultural Department has established. Appendix C.

cultural department before the peasants. The Department has been successful in introducing a new variety of cotton known as " American 4F " and two especially well adapted types of wheat, " Punjab 11 " and " Punjab 8A ". Other interests of the Department of Agriculture are to improve cattle-breeding, to provide information and demonstration farms for the development of fruit-growing, to aid the irrigation department in solving the growing problem of waterlogging and to educate the peasant in the necessity of applying fertilizer or a scientifically determined rotation of crops in order to insure the continued productivity of his fields. A new five year program of the Agricultural Department of the Punjab aims at placing a special officer in charge of each of the important crops for conducting research in every district. Each district is to have a farm under modern scientific management to be utilized as a model production and instruction centre as well as a seed distribution centre.[1] Thus the Government plans to educate the Punjab peasant to achieve for himself a higher standard of living. Progress has thus far been exceedingly slow, methods of working the land continue very much as they were before the British came, implements are still largely absent except in the rudest form, large-scale farming has not yet made its appearance and agriculture is still a primitive occupation carried on by petty peasant owners who are too seriously immersed in the problem of making a living to give much heed to advice, demonstration farms, literature which they cannot read or pure seed campaigns which they cannot understand. The Punjab peasant is doggedly sticking to the job of making a living with the methods which he has seen and tried, which have enabled his hardy race to continue to exist and which strangely suit his attitude toward life.

[1] *Department of Agriculture, Punjab Report*, 1928, *passim.*

The trend of this chapter has been to describe the village as the agricultural unit of the Punjab indicating that the village is still a far from luxurious or healthy place in which to live; that the holdings of land have been greatly subdivided and hence the productivity of the land has been somewhat lessened since efficient methods of production cannot well be used on plots so small and so scattered; that land values in the Punjab have risen steadily since the British began to govern this portion of India; that the price of land is continuing to increase in spite of the extension of the cultivated area; that the rise in price is probably a result of many causes among which must be mentioned the Punjab peasant's attitude toward land ownership as a sign of social prestige. It was further disclosed that there is no evidence to be derived from data available to indicate that the soil of the Punjab has been deteriorating during recent years. On the contrary, the produce per acre seems to have been increasing slightly since 1900. It has also been shown that production on irrigated land is from 30 to 100 per cent more profitable than on unirrigated land, measured in terms of produce, but that the increase in the value of land is not sufficiently explained by the increased production on irrigated land, while the crops produced on the irrigated areas in 1926-27 failed to support the belief that production on such land was particularly high. It was shown on the contrary that even though production on irrigated land is higher than on non-irrigated, the irrigated land is not proving as productive as might be expected, on account of the primitive methods employed by the peasant; that Government is not satisfied with the present state of agriculture in the province and is attempting to interest the peasant in a program of betterment including the use and selection of better seeds, rotation of crops, the use of better implements and the consolidation of scattered holdings. While some progress has been made, the peasant still follows methods which his forebears had used long before the British reached India and hence the program of changing the agricultural methods of the Punjab give small scope for optimism on the basis of solid achievement up to date. In the concluding chapter the problems relating to the cost of irrigation will be discussed.

CHAPTER VII

FINANCIAL ASPECTS OF CANAL IRRIGATION IN THE PUNJAB

AT the close of the fiscal year 1926-27 the Punjab was being served by a complex network of British planned, constructed and operated irrigation canals which irrigated 11,-157,624 acres of agricultural land. The length of main canals and branch channels in operation was 3,653 miles. Distributary channels carrying the water from the main canals to the village areas to be irrigated amounted to 13,-382 miles. Water courses carrying the much-needed moisture to the particular fields at the desire of the peasant must have presented an even larger total mileage.[1] It is thus apparent that the British in the Punjab have continued to expand and develop the irrigation plans which they inaugurated in 1849 when they annexed the province. This expansion of irrigation canals would have been far less effective had not the British projected a program of railway construction to complement the irrigation developments. In 1855, while outlining plans for further development of irrigation projects, officials responsible for such construction work wrote:

But the expediency of multiplying permanent canals of magnitude is doubtful until means of exporting the surplus produce shall have been provided. Until this cardinal and crying want, namely means of transportation, shall have been supplied, a great number of great canals would be in advance of the needs of the country.[2]

---

[1] *Irrigation Department Report, op. cit.*, part i, p. iv.
[2] *Administration Report, Punjab Government*, 1854-55, p. 61.

IRRIGATION, ITS FINANCIAL ASPECTS

A glance at the map showing the existing irrigation canals of the province and the map indicating the railways of the Punjab will show how closely the development of the two are inter-related.[1] A careful reading of the *Administration Reports of the Punjab Government* which have been issued annually since 1849-50 supports the view that the British have looked upon canal irrigation in the Punjab as a great commercial venture. The reports are replete with estimates of costs of constructing new canals and optimistic estimates of prabable earnings of the canals planned.

COST OF CONSTRUCTION OF IRRIGATION CANALS

The total capital expenditure of the Government on account of canal irrigation development from 1849 to 1927 in the Punjab amounted to Rs. 29,54,81,745. This sum includes expenditures on projects under construction in 1926-27. The importance of this point will be stressed later when the return on the Government's capital investment is shown. It will be shown that the capital expenditure of Rs. 6 crores on the Sutlej Valley Project is at present not earning a return on this sum, but in spite of this fact, the profits on the total capital invested to date for the year 1926-27 amounts to 14.38 per cent on productive works. In Table 15 the items composing the capital outlay as represented by the canals for which capital accounts are kept are presented.

The capital outlay of the British has been considerable as Table 15 indicates. The nature of the investment and the political problems involved in the construction of the canal system made it practically impossible for such a development to be successfully carried to fruition under any except governmental auspices. Previous chapters showed that the

[1] Maps on page 5 and 48. See also Appendix D and Appendix E giving statistics and date of development.

124   CANAL IRRIGATION IN THE PUNJAB

Government financed the development of canal-irrigation works out of current revenues up to 1882. After that date

TABLE 15. TOTAL CAPITAL OUTLAY ON PRODUCTIVE AND UNPRODUCTIVE CANAL IRRIGATION WORKS OF THE PUNJAB, WITH THE EXPENDITURE ON CAPITAL ACCOUNT FOR EACH CANAL 1849-1927 [1]

Productive works	Capital outlay
(1) Western Jumna Canal	Rs. 1,89,26,711
(2) Sirhind Canal	2,63,46,465
(3) Upper Bari Doab Canal	2,16,23,934
(4) Lower Bari Doab Canal	2,21,15,228
(5) Upper Chenab Canal	3,73,05,466
(6) Lower Chenab Canal	3,59,14,603
(7) Upper Jhelum Canal	4,42,56,719
(8) Lower Jhelum Canal	1,90,17,322
(9) Sidhnai Canal	13,29,569
(10) Muzaffargarh Inundation	18,91,493
(11) Chenab Inundation canals	11,63,323
(12) Sutlej Valley Project [2]	6,07,75,273
Central Workshop	8,82,313
Total	29,15,48,419
Unproductive works	
Shahpur Canal	2,25,925
Ghaggar Canal	3,88,435
Indus Canals	33,18,966
Total	39,33,326
Grand total capital outlay on irrigation canals	29,54,81,745 [3]

[1] Data for table taken from 1926-27, *Irrigation Department Report*, *op. cit.*, part ii, p. 8.

[2] The Sutlej Valley Project includes the old Upper Sutlej Inundation Canals.

[3] At the prevailing rate of exchange the Government's total capital investment in canal irrigation works in the Punjab would have amounted to $105,782,464.71 by 1927.

it was possible to finance irrigation projects which promised a fair return on the investment within a reasonable period of time from loans. It will be remembered that the period following 1882 was characterized as the period of rapid expansion of irrigation projects. The facility of the Government in securing funds from abroad for the construction of these ventures which promised such good returns on the capital invested is one of the chief causes for the speeding up of canal irrigation expansion during the latter period. A comparison of the figures presented in Table 15 with the estimated cost of constructing the various canals named will show that the estimated cost was in almost every case exceeded before the project was fully in operation. But in spite of this, the funds thus invested have proved a good income-producing source for the Punjab Government.

The important sources of income are water rates assessed according to area and crop irrigated, navigation fees collected from users of the canals for transportation purposes, sale of waste lands, increased land revenue and the sale of produce from plantations along the canals operated and controlled by the Government. The water rates are collected generally from the occupier or tenant of the land without effort to locate for collection of this fee, the owner of the particular plot on which water rates are assessed.[1] In some sections of the province the various charges due to Government, including land revenue, water rates and " land revenue due to irrigation " are paid jointly by owner and tenant. Since the province is so predominantly a land of small peasant proprietors the distinction as to whether owner or tenant pays the water rate is not important for purposes of this monograph. The land revenue in the Punjab is assessed at a flat rate per cultivated acre. In 1926-27 that rate was Rs. 1-8-0

[1] However, as has been indicated, the occupier and owner are generally identical.

per acre. But the irrigated land served by government canals is assessed in addition to the straight land revenue and the water rate by an additional amount called " land revenue due to irrigation ". This assessment first appeared in the accounts of the Government in the Punjab in 1869-70, and was thus described by the officials in their report:

The item " water advantage revenue " appears for the first time in the accounts and occurs in the districts affected by canals which have recently come under revision of Settlement, viz., Gurdaspore, Amritsar, Lahore and Montgomery. These districts were settled on the principle that the regular land revenue should be calculated at unirrigated rates and that the Government share of the increased productiveness of lands irrigated by canals should be assessed at a certain rate per acre, varying according to the locality. This rate is quite distinct from the water rate proper, imposed by the canal authorities as the price of the water supplied for irrigation; thus all canal irrigated lands have to pay three rates of assessment, viz., (1) the regular land revenue; (2) water advantage revenue and (3) the canal water rate.[1]

Since its first assessment, this " water advantage revenue " item has come in for a large share of criticism. It is difficult to discover the basis for this third rate, since it would appear to fall naturally under one of the other two assessments against the land, namely the water rate proper or the land revenue. Nor is it easy to discover the mode of assessing this additional rate on land. Since the revenue assessments are paid to the Government regardless of the name applied to the particular assessment it would seem as though the British have in this case merely added to their bookkeeping task an additional item. It could be quite logically included in the water rate since it applies only to lands irri-

[1] *Administration Report Punjab Government*, 1869-70, paragraph 100.

gated by canals. It might equally well be added to the land revenue proper, although in that case the flat-rate policy would apparently need to be discarded. In any case, this item,[1] " water advantage revenue " or " advantage due to irrigation " adds to the Government's income from irrigation projects in operation and as such is included in this monograph among the forms of income accruing to the Government on account of its control of irrigation canals in the Punjab.[2] Navigation fees and revenue from the sale of canal plantation products have not yielded any considerable sums to Government as will be indicated shortly. A source of revenue which promises to become more important in the future is that relating to the sale of water rights for the production of water-power along the canal rapids in some locations. Thus far this item has scarcely been developed, though, as will be remembered, it was emphasized in the estimates of the engineers who proposed and worked out plans for irrigating the Punjab.

[1] The irrigation receipts are swelled by an item called " portion of land revenue due to irrigation." A certain portion of the land revenue is said to be indirectly due to irrigation works constructed by the State. An attempt is made to credit the Irrigation Account with this portion of the Land Revenue. In the Irrigation Account we have the direct receipts which are derived from water rates paid for the use of water. In addition to these, we have this item, " portion of land revenue due to irrigation." . . . This is undoubtedly an arbitrary deduction from Land Revenue, the effect of which is to introduce an unnatural element in the accounts. The Land Revenue is shown at a smaller figure and the irrigation receipts are swelled. On the one hand the Government can say that the increase in Land Revenue, that is, the burden on the agriculturist is smaller than it really is; on the other hand they can show that their undertakings in the Irrigation Department yield larger receipts and are hence very successful.—Vakil, C. N., *Financial Developments In Modern India* (Bombay, 1925), p. 245.

[2] This portion of land revenue due to irrigation introduces an element of uncertainty into government's taxation policy. The peasant is at a loss in attempting to estimate the amount of this portion of his taxes.

The importance to the Punjab Government of receipts from irrigation services will become evident from a brief study of the 1927-28 budget estimate as to revenue. The total receipts of the Government from all sources were estimated at Rs. 15,79,80,000 of which Rs. 4,67,42,000 were to be received from the irrigation department. This estimate of 1927-28 receipts was designated as Net Irrigation Receipts. The expenses of administration of the irrigation services had been deducted from the final estimate above. Income taxes were estimated to produce Rs. 4,23,000 or less than 10 per cent of what was expected from irrigation. Net land revenue was estimated at Rs. 2,84,63,000 lagging far behind the Government's estimate of the irrigation revenues. Thus for 1927-28 the irrigation receipts were to provide almost one-third of the total income of the Punjab Government. Irrigation has thus become by far the most important single source of revenue for the Punjab Government. The estimate for 1927-28 for irrigation receipts of Rs. 4,67,-42,000 [1] compares with an actual assessed revenue received of Rs. 4,19,23,010 [2] in 1926-27. The difference is to be accounted for in the increased area which was to have been brought under irrigation in the latter year due to the opening of portions of the Sutlej Valley Project and to the expectation that 1927-28 would be a more favorable year for cotton producers, making it unnecessary for the Government to remit as large an amount of water rates to peasants whose crops had not matured as it had in 1926-27.[3] This brief

[1] *Indian Yearbook* (Bombay, 1928), p. 136.

[2] *Irrigation Report, op. cit.*, part ii, pp. 8-11.

[3] The Government practically guarantees the peasant a crop on his irrigated land. If for any reason the crop fails to mature, the Government remits the water rate. Hence in 1926-27 the area on which irrigation charges were remitted was no less than 1,083,896 acres; of this 87,160 acres represented remissions "under the terms of colonization" or in

statement of the budget estimate of the Punjab Government for 1927-28 indicates the importance of irrigation receipts which flow into the general coffers of the State. Without this source of income it seems that the Government would necessarily have to curtail its expenditures severely or increase its tax rates on other taxable items or discover new sources for levying taxes. We shall revert to this aspect of the problem later in this chapter.

In Table 15 the canals are numbered from one to twelve according to their position in the table. The same numbers are used in the section immediately following so that reference to Table 15 may be easily made. In the tables which follow are given data relating to the area irrigated by each canal in 1926-27, the amount collected in the form of water rates, from those who benefited by the irrigation service, receipts from plantations, water-power fees, navigation and other charges. The tables also indicate the amount collected under the heading " indirect receipts " which relates to the portion of the land revenue due to irrigation. Finally each table indicates the working expenses of the year in relation to each canal designated, the net gain or loss, and the percentage of gain or loss in relation to the total capital investment of each canal project. The canals now to be considered as to

respect of crops sown in the moisture of a previous crop; the area of crops that failed either wholly or partially, therefore, amounted to 996,736 acres, or 8.93 per cent. of the area sown. . . . The remissions of 379,992 acres granted under the special orders of the Governor in Council on account of the general failure of the cotton crops is solely responsible for the increase in the area remitted.—*Irrigation Report, op. cit.*, pt. i, p. 3.

Of the remaining 616,744 acres remitted under ordinary rules, 89,219 acres were remitted on account of a short supply of water, 1,570 acres on account of damage done by hail, 24,171 acres on account of floods, 32,838 acres on account of faulty germination of seeds, 110,324 acres on account of damage done by locusts, other insects and blight and on 128,729 acres for miscellaneous reasons not indicated in the report of the Irrigation Department.

their financial position are the Western Jumna, Sirhind, Upper Bari Doab (originally called the New Baree Doab), Lower Bari Doab, Upper Chenab, Lower Chenab, Upper Jhelum, Lower Jhelum, Sidhnai, Muzaffargarh Inundation, Chenab Inundation and the Sutlej Valley Project. This completes the list of " productive canals " of the Punjab. This section closes with a discussion concerning the financial aspects of the " unproductive canals " included in Table 15, and a brief reference to the Central Work Shop.

(1) *The Western Jumna Canal*

This canal as will be remembered from previous chapters was taken over by the British as a going concern about the middle of the 19th century. Hence, as indicated in Table 15 *supra* a large portion of its capital cost had been met by previous rulers of the area served by the Western Jumna Inundation canals so that the British capital outlay was not as great as might have been expected in consideration of the area irrigated. The canal was a productive work from its initial operation under the British. Hence, no " arrears of interest " item ever occurs in its financial history. By 1901 the entire capital expenditure which the British had made in extending the scope of this project had been repaid. By 1926-27 the accumulated surplus revenues of the Western Jumna Canal stood at Rs. 6,69,31,107.[2] Since the total capital outlay of the British had been but Rs. 1,89,26,711[3] it is evident that the canal had been paid for no less than four times out of its earnings. Table 16 indicates its financial status on the basis of operation during 1926-27.

[1] *Irrigation Department Report*, 1926-27, *op. cit.*, pt. i, p. 9.
[2] *Ibid.*, pt. ii, p. 8.
[3] *Ibid.*, pt. ii, p. 8.

IRRIGATION, ITS FINANCIAL ASPECTS

TABLE 16. FINANCIAL RECORD OF WESTERN JUMNA CANAL IN 1926-27 [1]

Total area irrigated and assessed to water rates ....825,669 acres	
Revenue assessed during the year:	
Occupier's rates....................................	Rs. 37,19,061
Plantation assessed receipts........................	17,036
Water power..	7,610
Navigation receipts.................................	1,80,845
Miscellaneous receipts..............................	58,192
Total direct assessed receipts	39,82,744
Indirect receipts [2]...............................	2,26,417
Grand total assessed receipts..................	42,09,161
Working expenses for year........................	17,76,309
Net assessed revenue	Rs. 24,32,852
Percentage earned on total capital outlay..........	12.85,%

(2) The Sirhind Canal

The Sirhind Canal project was begun about 1870 and opened for irrigation service in 1883. This canal was not a financial success during the first decade of its operation because of the silting of the channel and the distributaries. These difficulties have since been successfully overcome. By 1921 the whole capital outlay of Rs. 2,63,46,465 [3] had been repaid out of earnings. From that year no interest needed to be any longer paid on the capital that had been invested in the project. By 1927 the accumulated surplus revenues stood at Rs. 4,22,28,099 [4] almost double the total capital outlay incurred in its construction. The Sirhind Canal in 1926-27 irrigated a total of 1,659,323 acres of which 551,728 were in Native States' territory. Table 17 indicates its financial status on the basis of operation during 1926-27.

[1] *Irrigation Department Report*, 1926-27, *op. cit.*, pt. ii, pp. 8-11.
[2] 'portion of land revenue due to irrigation.'
[3] *Cf. supra*, Table 15; *Irrigation Report, op. cit.*, pt. i, p. 11.
[4] *Ibid.*, pt. ii, p. 8.

TABLE 17. FINANCIAL RECORD OF SIRHIND CANAL IN 1926-27 [1]

Total area irrigated and assessed to water rates in British Territory only during the year	1,107,595 acres
Revenue assessed during the year:	
Occupier's rates	Rs. 47,63,580
Plantation assessed receipts	30,414
Water power	1,42,515
Navigation receipts	10,766
Miscellaneous receipts	2,25,421
Total direct assessed receipts	51,72,696
Indirect receipts [2]	1,63,369
Grand total assessed receipts	53,36,065
Working expenses for the year	14,10,620
Net assessed revenue	39,25,445
Percentage earned on total capital outlay	14.90%

(3) *The Upper Bari Doab Canal*

As will be remembered from the discussion in Chapter II, the New Baree Doab, or as it is now designated, the Upper Bari Doab Canal was the first extensive new project in canal irrigation to be constructed by the British in the Punjab. It was originally opened for service in 1859 but it was not completed in its present scope until 1879. Since 1889 this canal has been one of the most successful of the irrigation projects in the Punjab. By 1905 the total capital investment plus arrears of interest of Rs. 2,16,23,934 was repaid out of earnings.[3] By 1927 the accumulated surplus revenue account stood at Rs. 7,96,03,356, indicating that the canal had paid for itself more than three times out of earnings to that date. In Table 18 the financial results of its operation during 1926-27 are shown. The net earnings for that year were 19.84 per cent on the capital investment which the project represents.

---

[1] *Ibid.*, pt. ii, pp. 8-11.
[2] 'portion of land revenue due to irrigation.'
[3] *Irrigation report*, 1926-27, *op. cit.*, part i, p. 14.

TABLE 18. FINANCIAL RECORD OF THE UPPER BARI DOAB CANAL IN 1926-27 [1]

Total area irrigated and assessed to water rates	1,276,495 acres
Revenue assessed during the year:	
Occupier's rates	Rs. 50,73,949
Plantation assessed receipts	17,762
Navigation receipts	
Water power	1,00,712
Miscellaneous receipts	1,01,847
Total direct assessed receipts	52,94,270
Indirect receipts [2]	10,34,635
Grand total assessed receipts	63,28,905
Working expenses for the year	20,36,948
Net assessed revenue	Rs. 42,91,957
Percentage earned on capital outlay 1926-27	19.84%

(4) The Lower Bari Doab Canal

As was shown in Chapter IV the Lower Bari Doab Canal is one of the Triple Canal Project features. This portion of the larger project was begun in 1907 and opened for irrigation in 1913. It was not until 1917, however, that it was entirely completed. The capital outlay on the Lower Bari Doab was Rs. 2,21,15,228 [4] and so profitable has the operation of this canal proved to be that the total capital investment was repaid out of earnings by 1926 [5] while its accumulated surplus revenues stood at Rs. 3,14,39,639 at the end of the fiscal year 1926-27. [6] As indicated in Table 19, the net

[1] *Irrigation Report*, 1926-27, *op. cit.*, part ii, pp. 8-11.

[2] 'portion of land revenue due to irrigation.'

[3] The item net assessed revenue represents a net return from the particular irrigation project which is the subject of this series of tables, 16 to 28. This surplus over working expenses while entered in the accounts as "Accumulated surplus revenues" is not kept as a separate fund but goes into the general revenues of the Punjab Government. But this does not lessen the importance of the net return from irrigation due to the operation of the irrigation canals.

[4] *Cf. supra*, Table 15.

[5] *Irrigation Report*, 1926-27, *op. cit.*, p. 18.

[6] *Ibid.*, part ii, p. 8.

assessed revenue for that year was 32.15 per cent on the capital outlay. Its financial position is shown in the table.

TABLE 19. FINANCIAL STATUS OF THE LOWER BARI DOAB CANAL IN 1926-27 [1]

Total area irrigated and assessed to water rates	1,220,034 acres
Revenue assessed during the year:	
Occupier's rates	Rs. 41,60,125
Plantation assessed receipts	3,852
Water power	8,035
Navigation receipts	
Miscellaneous receipts	1,99,573
Total direct assessed receipts	43,71,585
Indirect receipts [2]	43,52,830 [3]
Grand total assessed receipts	87,24,415
Working expenses for the year	16,13,740
Net assessed revenue	Rs. 71,10,675
Percentage earned on capital outlay	32.15%

(5) *The Upper Chenab Canal*

The Upper Chenab Canal is also a part of the Triple Canal Project. This portion of the project was begun in 1905, first opened for irrigation service in 1912 and entirely completed in 1917. The capital outlay necessary to complete this canal was Rs. 3,73,05,466 and all arrears of interest were entirely paid out of earnings by 1925.[1] Since 1925 the canal has not been able to meet its interest charges on the capital investment with the result that an accumulated interest charge of Rs. 15,22,378 has developed. Thus none of the capital cost of this canal has yet been repaid out of earnings. However, it would be illogical

---

[1] *Irrigation Report*, 1926-27, *op. cit.*, part ii, pp. 8-11.

[2] 'portion of land revenue due to irrigation.'

[3] Of the Rs. 43,52,830 designated in the table as 'indirect receipts' Rs. 38,74,483 came under the heading 'portion of land revenue due to irrigation' while the remainder represents sales of Crown lands in the Lower Bari Doab Canal tract during the year. *Cf. ibid.*, part ii, p. 6.

[4] *Irrigation Report, op. cit.*, p. 20.

IRRIGATION, ITS FINANCIAL ASPECTS 135

to conclude that the Upper Chenab Canal is not a paying venture on this account. It will be remembered that this canal is a part of the larger Triple Canal Project, that its chief function is to carry the waters of the Chenab River at Merala to the River Ravi at Balloki where the Lower Bari Doab, the third part of the triple project takes off. In Table 19 *supra* the profitable nature of the Lower Bari Doab Canal was indicated. Since the two canals are parts of the same project and since one feeds the other it is somewhat fictitious to consider the financial record of each alone. Considered as a unit the Upper Chenab Canal's operation costs plus the " portion of land revenue due to irrigation " during 1926-27 were greater than its earnings. But the loss is more apparent than real as will be shown following the presentation of its financial record for the year 1926-27.

TABLE 20. FINANCIAL RECORD OF THE UPPER CHENAB CANAL IN 1926-27 [1]

Total area irrigated and assessed to water rates	501,210 acres
Revenue assessed during the year:	
Occupier's rates	Rs. 24,79,765
Plantation assessed receipts	1,417
Water power	
Navigation receipts	
Miscellaneous	22,816
Total direct assessed receipts	25,03,998
Indirect receipts [2]	20,96,826 [3]
Grand total assessed receipts	4,07,172
Working expenses for the year	24,44,578
Net assessed revenue	—20,37,406
Percentage loss on capital outlay for 1926-27	5.46%

[1] *Irrigation Report*, 1926-27, *op. cit.*, pp. 8-11.

[2] 'Portion of land revenue due to irrigation.'

[3] It will have been noted that this item of indirect receipts is a minus quantity in this table. The method of calculating the indirect receipts when they are not collected is not indicated in the reports. It does not seem logical to assess this item in the case of this canal since the advantage appears to have been less than the amount assessed would indicate. It will be noted also that the loss on the operation of this canal during the year was just a little less than the indirect assessment. Hence,

(6) The Lower Chenab Canal

As has been shown, the Lower Chenab Canal was opened as an inundation canal in 1887 and completed as a perennial canal in 1892. The capital outlay represented by this canal

TABLE 21. FINANCIAL RECORD OF THE LOWER CHENAB CANAL IN 1926-27 [1]

Total area irrigated and assessed to water rates.	2,562,136 acres
Revenue assessed during the year:	
Occupier's rates	Rs. 1,17,59,871
Plantation assessed receipts	19,028
Water power	27,257
Navigation	
Miscellaneous	1,30,340
Total direct assessed receipts	1,19,36,496
Indirect receipts [2]	1,16,74,094
Grand total assessed receipts	2,36,10,590
Working expenses for the year	31,58,982
Net assessed revenue	Rs. 2,04,51,608
Percentage earned on capital outlay	56.95%

the loss of 5.46 per cent on the capital investment represented by this canal is a fictitious loss appearing only on the books. If only the direct assessed receipts and the working expenses are considered a profit of slightly less than Rs. 1 lakh would be shown. When the operation of the Upper Chenab Canal is considered in relation to the total operation of the Triple Canal Project of which it forms a part, the earnings of the other two sections of the work are sufficient to overbalance this apparent deficit of this portion of the project. To make this report comparable to those of the other canals for which tables are given in this monograph the financial records as indicated in Table 20 are presented. The figures are taken from the report of the Irrigation Department of the Punjab Government.

[1] *Irrigation Report*, 1926-27, *op. cit.*, part ii, pp. 8-11.

[2] 'Portion of land revenue due to irrigation.' From the discussion of this canal in Chapter IV it will be remembered that the area served by this canal was prior to irrigation waste land. Thus it is evident that any land revenue collected from that portion of the irrigated area which had been worthless waste before the canal was built would logically fall under the heading 'portion of land revenue due to irrigation.' Land revenue was practically non-existent in this area prior to the construction of this canal and the colonization of the tract thus irrigated.

IRRIGATION, ITS FINANCIAL ASPECTS

is Rs. 3,59,14,603;[1] by 1899 all arrears of interest on capital had been paid out of earnings, thus conclusively placing the Lower Chenab in the class of productive works. By 1906 the whole capital was reimbursed out of earnings[2] and by 1927 the accumulated surplus revenue account for this canal stood at Rs. 27,62,64,925 or roughly eight times the original cost of the project.[3] As shown in Table 21 the net assessed revenue for 1926-27 amounted to 56.95 per cent on the capital cost of the canal. Its financial record is presented in Table 21.

(7) *The Upper Jhelum Canal*

The Upper Jhelum Canal is also one of the canals comprising the Triple Canal Project. Construction of this

TABLE 22. FINANCIAL RECORD OF THE UPPER JHELUM CANAL IN 1926-27 [4]

Total area irrigated and assessed to water rates.........	302,707 acres
Revenue assessed during the year:	
Occupier's rates................................	Rs. 12,85,982
Plantation assessed receipts.....................	983
Water power	
Navigation	
Miscellaneous	67,662
Total direct assessed receipts	13,54,627
Indirect receipts[5]	2,51,098
Grand total assessed receipts	16,05,725
Working expenses	12,27,949
Net assessed revenue	Rs. 3,77,776
Percentage earned on capital outlay.........	0.85%

canal was commenced in 1905, it was opened for irrigation service in 1915 and completed in its present form in 1917. The total capital outlay incurred in its construction was Rs.

[1] *Cf. supra*, Table 15.
[2] *Irrigation Report*, 1926-27, *op. cit.*, p. 22.
[3] *Ibid.*, part ii, p. 8.
[4] *Irrigation Report*, 1926-27, *op. cit.*, pt. ii, pp. 8-11.
[5] 'Portion of land revenue due to irrigation.'

4,42,56,719 [1] and the arrears of interest on the capital invested stood at Rs. 2,04,53,289 in 1926-27.[2] This canal is chiefly a feeder and the irrigation of the area through which it flows is of secondary importance. It carries the water of the Jhelum River to the Chenab River whence it is carried through the Lower Chenab Canal and its distributaries to irrigate over two and a half million acres of land as has been shown in the section immediately preceding. It has been shown that the Lower Chenab Canal earned 56.95 per cent on its capital cost in 1926-27. The Lower Chenab however depends for its waters upon the Upper Jhelum Canal. Since the revenue from each of the canals flows into the coffers of the same government it would be useless and pedantic to attempt to designate the portion of the earnings of the Lower Chenab which is due to the cooperation of the Upper Jhelum Canal which feeds it. Table 22 shows the results of the operation of this canal during 1926-27.

(8) *The Lower Jhelum Canal*

This canal was completed in its present scope in 1917 although it was opened for irrigation service in 1901. The development of this project greatly aided the economic position of the districts of Shahpur, Gujrat and Jhang which are served by the Lower Jhelum Canal. The total capital outlay incurred in its construction was Rs. 1,90,17,322 as was shown in Table 15. During the first five years of its operation no irrigation revenues accrued from the assessment of water rates due to a policy of the Government of encouraging the development of the tract irrigated. That the canal has been a profitable investment for the Punjab Government is evident from the fact that by 1917 the entire capital outlay had been reimbursed out of earnings from assessments and

---

[1] *Cf. supra*, Table 15.
[2] *Irrigation Report, op. cit.*, pt. ii, p. 8.

IRRIGATION, ITS FINANCIAL ASPECTS

sales of waste lands.[1] By 1927 the total accumulated surplus revenues were Rs. 4,82,67,990 indicating that the canal had paid for its construction several times out of its earnings. The financial results of its operation in 1926-27 are indicated in Table 23 which shows a net revenue accruing of Rs. 43,-53,708 or 22.89 per cent on its capital cost.

TABLE 23. FINANCIAL RECORD OF THE LOWER JHELUM CANAL IN 1926-27 [2]

Total area irrigated and assessed to water rates	881,081 acres
Revenue assessed during the year:	
Occupier's rates	Rs. 34,96,445
Plantation assessed receipts	2,581
Water power	38,798
Navigation	
Miscellaneous	37,162
Total direct assessed receipts	35,74,986
Indirect receipts[3]	22,55,181
Grand total assessed receipts	58,30,167
Working expenses for the year	14,76,459
Net assessed revenue for the year	Rs. 43,53,708
Percentage earned on capital outlay	22.89%

(9) *The Sidhnai Canal*

The Sidhnai Canal is one of the most important inundation canals of the Punjab. In the Sidhnai System are included the Koranga Canal, the Fazil Shah Canal and the Abdul Hakim Canal. The main canal takes off from the Ravi River just above its junction with the Chenab River. The inundation canals, since they are dependent for their water flow on the natural level of the rivers, are not as certain and regular in their irrigation service as are the perennial canals. However, to offset this disadvantage, the cost of constructing inundation canals is generally less than the outlay required for a perennial canal of similar proportions due to the absence of elaborate and costly headworks on the

---

[1] *Irrigation Report, op. cit.*, pt. i, p. 28.
[2] *Irrigation Report, op. cit.*, part ii, pp. 8-11.

former. The Sidhnai Canal was opened in 1886 after a capital cost of Rs. 13,29,569 had been incurred as shown in Table 15 *supra*. The capital outlay has been earned many times over as is indicated by the accumulated surplus revenue account of Rs. 1,19,34,266 [1] in 1926-27. The financial results of its operation during the year 1926-27 are shown in Table 24.

TABLE 24. FINANCIAL RECORD OF THE SIDHNAI CANAL IN 1926-27 [2]

Total area irrigated and assessed to water rates	332,489 acres
Revenue assessed during the year:	
Occupier's rates	Rs. 5,04,158
Plantation assessed receipts	1,978
Water power	
Navigation	
Miscellaneous	
Total direct assessed receipts	5,09,383
Indirect receipts [3]	5,39,982
Grand total assessed receipts	10,49,365
Working expenses for the year	1,15,978
Net assessed revenue for the year	Rs. 9,33,387
Percentage on outlay of capital	70.20%

(10) *The Muzaffargarh Inundation Canals*

The Muzaffargarh Inundation Canal System includes the series of small canals taking their water from the Indus and Chenab Rivers in the fork of these two rivers. The canals included in the system are the Kot Sultan, Hazara, Magasson, Magi, Ghuttu, Puran, Suleman, Karam, Genesh and the Taliri. These canals represent a British capital investment of Rs. 18,91,493 as was shown in Table 15. The year 1926-27 was an abnormal year for this system of canals because of the high working expenses which were necessary on account of certain improvements which were made involving the removal of silt from the channels. Although 1926-27

---

[1] *Irrigation Report*, 1926-27, *op. cit.*, part ii, p. 8.

[2] *Ibid.*, part ii, pp. 8-11.

[3] 'Portion of land revenue due to irrigation.'

IRRIGATION, ITS FINANCIAL ASPECTS

was an unprofitable year for this system of canals, closing with a net loss of Rs. 14,710 or .77% on the capital invested, this is but a temporary situation as is proved by the total "surplus revenues" account which has reached the sum of Rs. 98,32,107 approximately five times the total capital cost.[1] Table 25 shows the financial results for the year 1926-27.

TABLE 25. FINANCIAL RECORD OF THE MUZAFFARGARH INUNDATION CANALS IN 1926-27 [2]

Total area irrigated and assessed to water rates	335,097 acres
Revenue assessed during the year:	
Occupier's rates	Rs. 2,18,105
Plantation assessed receipts	22,224
Water power	
Navigation	
Miscellaneous	4,150
Total direct assessed receipts	2,44,479
Indirect receipts [3]	3,21,306
Grand total assessed receipts	5,65,785
Working expenses for the year	5,80,495
Net assessed revenue for the year	Rs.—14,710
Percentage loss on capital outlay in 1926-27	0.77%

(11) *The Chenab Inundation Canals*

The group of inundation canals comprising this system was completed in 1895. The canals making up the Chenab Inundation Canal system, the Mattital, Daurana Laugana, Wali Mohammad, Sikandarabad, Gajjhuhatta and the Balochwah, combine to offer irrigation facilities to some 212,404 acres though a network of channels aggregating 221 miles.

[1] *Irrigation Report*, 1926-27, op. cit., part ii, p. 8.

[2] *Ibid.*, part ii, pp. 8-11.

[3] 'Portion of land revenue due to irrigation'. It will be noted that the indirect receipts on the Muzaffargarh Inundation canals are larger than the total direct assessed receipts on account of irrigation service. When it is remembered that the area irrigated by this system of inundation canals lies in that portion of the Punjab in which the normal average rainfall per year is less than five inches it will be realized that whatever produce is made possible in this area is due to irrigation. Hence, practically the whole land revenue collected in this region is due to irrigation.

The total capital outlay on this system amounts to Rs. 11,-63,323 and the "accumulated surplus revenues" account at the end of 1926-27 stood at Rs. 1,07,01,903,[1] indicating that the canals of this group had repaid the capital cost of their construction about nine times. This "accumulated surplus revenues" account is an important index of the profitableness of the government canals of the Punjab. In the accompanying table the financial results of the operation of this system of canals in 1926-27 is shown.

TABLE 26. FINANCIAL RECORD OF THE CHENAB INUNDATION CANALS IN 1926-27 [2]

Total area irrigated and assessed to water rates	212,404 acres
Revenue assessed during the year:	
Occupier's rates	Rs. 2,66,930
Plantation assessed receipts	904
Water power	
Navigation	
Miscellaneous	903
Total direct assessed receipts	2,68,737
Indirect receipts [3]	2,89,970
Grand total assessed receipts	5,58,707
Working expenses for the year	2,12,557
Net assessed revenue for the year	Rs. 3,46,150
Percentage on capital outlay earned in 1926-27	29.77%

(12) *The Sutlej Valley Project*

As shown in Chapter IV the Sutlej Valley Project is in process of construction. However, one branch of the new works was opened to irrigation during the year 1926-27, namely, the Pakpattan Canal, and the old Upper Sutlej Inundation canals continued to operate though they had been amalgamated with the larger project.

This is the fifth year of extensive work on the Sutlej Valley

---

[1] *Irrigation Report*, 1926-27, op. cit., part ii, p. 8.

[2] *Ibid.*, op. cit., pp. 8-11.

[3] 'Portion of land revenue due to irrigation.'

Project. The expenditure under capital incurred during the year amounted to Rs. 2,87,10,972 bringing the total outlay at the end of the year under review to Rs. 12,18,89,274 including Rs. 6,11,14,001 contributed by the States of Bahawalpur and Bikaner, copartners in the Project, and Rs. 18,02,625 transferred from the Capital Account of the old Upper Sutlej Inundation Canals.[1]

The total capital outlay contributed by the British thus amounted to Rs. 6,07,75,273 to the end of the year 1926-27. The original estimates for the cost of construction of the whole project in British territory amounted to Rs. 5,58,-47,600 which have already been exceeded by approximately half a *crore* of rupees. When once completed this project is destined to irrigate not less than 3,600,329 acres in British territory. Of this only 26,256 acres were being irrigated by new works in 1926-27. The table which follows shows the financial status of the whole project but relates to the earnings of the Old Upper Sutlej Inundation Canals only.

TABLE 27. SUTLEJ VALLEY PROJECT FINANCIAL RECORD IN 1926-27 [2]

Total area irrigated and assessed to water rates	338,019 acres
Revenue assessed during the year:	
Occupier's rates	Rs. 5,57,649
Plantation assessed receipts	1,643
Water power	
Navigation	
Miscellaneous	6,497
Total direct assessed receipts	5,65,789
Indirect receipts [3]	2,65,581
Grand total assessed receipts	7,91,370
Working expenses for the year	9,73,656
Net assessed revenue for the year	Rs.—1,82,286
Percentage deficit on capital outlay in 1926-27	0.29%

Included in the productive irrigation works of the Punjab

[1] *Irrigation Report, op. cit.*, part i, p. 38.
[2] *Ibid.*, part ii, pp. 8-11.
[3] 'Portion of land revenue due to irrigation.'

are the Central Workshops located at Madhopor. These workshops were constructed soon after the irrigation projects of the British were planned and launched in the Punjab. Here most of the simple tools and machines used in the construction of the canals have been manufactured and repaired. The construction and development of the workshops during the period of canal construction since 1850 amounted to Rs. 8,82,313. While the workshops are productive in the sense that they make possible the more efficient construction and maintenance of the entire irrigation canal system of the province, it is not surprising that the workshops, considered separately, generally present a deficit at the end of the year's operation. Thus, in 1926-27, miscellaneous receipts of the workshops amounted to Rs. 2,21,936 while working expenses were Rs. 2,88,082, creating a net deficit for the year of Rs. 66,146 or 7.49 per cent on the total capital expended on these workshops. It is logical, however, to include the workshops account in the financial review of the productive irrigation works of the Punjab, since these workshops have been instrumental in aiding in the construction of all of the irrigation projects. Hence, in the financial record of the productive irrigation canals of the Punjab for the year 1926-27, the workshops account is included in the capital cost of the irrigation schemes. The table which follows combines the data which have been presented in the tables relating to the financial record of each of the productive works. The table is a summary of the data presented in the same manner as the individual tables have been built up. The total accumulated surplus revenues of the canals thus far reviewed amount to Rs. 58,27,01,089,[1] approximately twice as large a sum as the total capital cost of the productive irrigation canals combined.

Table 28 indicates that the collection of water rates on the areas irrigated by the productive irrigation works in the Pun-

---

[1] *Irrigation Report, op. cit.*, part ii, p. 8.

TABLE 28. FINANCIAL RECORD OF THE PRODUCTIVE IRRIGATION CANALS OF THE PUNJAB IN 1926-27, BEING A SUMMARY OF THE DATA IN TABLES 15-27 INCLUSIVE [1]

Total capital outlay to end of 1926–27	Rs. 29,15,48,419
Total accumulated surplus revenues	58,27,01,089
Total area irrigated and assessed to water rates	1,04,66,574 acres
Revenue assessed during the year:	
Occupier's rates	Rs. 3,82,85,620
Plantation	1,19,822
Water power	3,24,927
Navigation	1,91,611
Miscellaneous	10,79,746
Total direct assessed receipts	4,00,01,726
Indirect receipts [2]	1,92,37,637
Grand total assessed receipts	5,92,39,363
Working expenses for the year	1,73,16,353
Net assessed revenue receipts for year	Rs. 4,19,23,010
Percentage earned on total capital outlay in 1926–27, 14.38%	

jab is the most remunerative source of irrigation revenue; that the revenue derived from sales of produce from the canal plantations, water-power rights and navigation fees has thus far been unimportant since the combined revenue from these three minor sources in 1926-27 amount to less than 2 per cent of the total revenue derived from irrigation service; that the "portion of the land revenue due to irrigation" produces approximately one-third of the total irrigation revenue. This table indicates that the productive irrigation canals of the Punjab earned a net surplus of Rs. 4,19,23,010 or the equivalent of 14.38 per cent on the total capital investment in these canals. This percentage of earnings to capital investment would be considerably greater had not the Rs. 6,07,75,273 capital outlay of the incomplete Sutlej Valley Project been included in the total capital outlay. Thus the return on the capital investment in the future will probably be even larger than it has been in the past due to the earning power of the Sutlej Valley Project when once it is completed. Nor was

[1] *Irrigation Report, op. cit.,* part ii, pp. 8-11.

[2] 'Portion of land revenue due to irrigation.'

1926-27 a year of abnormally high earnings for the irrigation department as is indicated by the return on the capital outlay on productive works in previous years.[1] It is evident that the irrigation canals which have been surveyed in this section are even more productive than the figures given would indicate, since the works in operation are called upon to provide the costs out of current revenues collected from water rates and other irrigation services of new works projected. Finally it is evident from this study that the irrigation works have been successfully paid for out of their earnings and these irrigation department earnings have been collected from the people of the province.

In addition to the productive works which have been reviewed, there are a number of so-called " unproductive works " which include the Shahpur Canal, the Ghaggar Canal and the Indus Inundation Canals. These three canals combined to irrigate 327,919 acres during 1926-27[2] and represent a capital outlay of Rs. 39,33,326.[3] Direct and indirect assessed receipts on these canals amounted in that year to Rs. 8,80,661 while their costs of operation totaled Rs. 8,11,500 thus leaving a net revenue of Rs. 69,161 to be applied to general revenues of the Government. This amounted to 1.76 per cent on the capital investment incurred in their construction. These canals were constructed by the Government chiefly for purposes of famine prevention. They are therefore a kind of insurance of the British against the hazard of famine in the areas irrigated thereby. They are thus chiefly protective works and are not to be confused with the productive works which have been built with the definite aim of securing revenue for the Government. Not only have these unpro-

---

[1] *Cf. infra*, Table showing revenue of productive canal works 1887-1927 in Appendix H.

[2] *Irrigation Report, op. cit.*, p. 8.

[3] *Ibid.*, part ii, pp. 8-11.

ductive canals here referred to prevent famine, but they have added slightly to the general revenues of the Punjab Government.

One other minor canal should be reviewed briefly, namely, the Lower Sutlej Inundation Canal. This canal was in operation when the British took possession of the Punjab. Certain extensions and improvements of a minor nature have been made since that time, but the capital outlay of the British on this canal has not been large enough to justify keeping capital accounts for the project. Though the capital outlay is not indicated in the reports of the Irrigation Department of the Punjab Government, its gross receipts and its gross working expenses are given. Thus, in 1926-27 the Lower Sutlej Inundation Canal irrigated an area of 339,982 acres, its gross receipts were Rs. 7,74,370 while its working expenses were Rs. 2,20,094 leaving a net revenue of Rs. 5,-54,276.[1] What percentage this net earning was in relation to the capital outlay it is impossible to discern from a careful study of the reports. Since no mention of extensive developments in this area is made in the documents, relating to the development of irrigation canals in the Punjab, it is probable that the British have never incurred any large expense in this canal and hence the net revenue is all the more acceptable coming as it does from a canal which was operative when they conquered the Punjab. It is not at all improbable that the combined earnings of the so-called "unproductive canals" including this Lower Sutlej Inundation Canal would compare quite favorably with the earnings of the "productive works" provided only British capital outlay were used as the basis of estimating the percentage of profit each year.

The following statement, made in 1921 by officials of the Government of the Punjab, is even more applicable to the irrigation situation in the province today:

[1] *Irrigation Report, op. cit.*, part ii, p. 8.

The profits realized from irrigation give some indication of the advantages which accrue to an economically virgin country from the scientific development of its resources. This is absolute profit, after deducting the interest on the loans raised. . . . It would be difficult to find such a profitable investment elsewhere. This of course leaves out of account the other revenue the Government derives from the Canal Colonies, all ultimately due to the Irrigation Department. And the direct receipts to Government are but a fraction of the total increase of wealth to the people of the Punjab.[1]

The data presented and briefly discussed in the preceding section of this chapter conclusively support the thesis that canal irrigation development on the part of the Punjab Government has been a profitable venture. The Government's return on its capital investment has been quite satisfactory. The total capital outlay has been returned twice over in the earnings of the Irrigation Department. It would indeed be difficult to find such a profitable investment elsewhere. In the closing section of this chapter the cost of irrigation from the point of view of the Punjab peasant is considered. In as much as the peasant's contact with government in relation to irrigation centers about the payment and assessment of water rates, that is used as the introduction to a review of the peasant's burden as represented by irrigation costs.

THE COST OF CANAL IRRIGATION TO THE PUNJAB PEASANT

In the final analysis of the financial aspects of irrigation in the Punjab it is evident that the peasant who works upon the irrigated land served by government irrigation canals is the ultimate source of profit. The peasant pays the water rates, the portion of the land revenue due to irrigation and in large measure the minor shares of the Government revenue due to sales of plantation produce from canal plantations, naviga-

---

[1] *Administration Report*, 1921-22, *op. cit.*, p. 74.

IRRIGATION, ITS FINANCIAL ASPECTS

tion fees on haulage of the agricultural produce on the canals and the miscellaneous assessments on account of irrigation. It is important to study the incidence of the cost of irrigation upon the peasant.

As has been shown, water rates constitute the chief item among the various classes of canal-irrigation revenue. These rates are assessed at a flat rate per acre per crop matured in any given area. The rate may differ, however, from district to district. Thus the water rates are not collected on a flat acreage basis or according to any accurate measure of quantity of water supplied. However, the rates per crop differ roughly in relation to the amount of water the various crops require. Thus the water rate per acre of sugarcane matured in the Upper Bari Doab Canal area is Rs. 12-1-7 while the water rate per acre of wheat matured in the same area is Rs. 5-4-8. Let us examine the economic position of a peasant in the canal colonies occupying an irrigated plot of an average size, namely 9.8 acres and discover as accurately as possible on the basis of the British estimates of normal crops and the water rates, land revenue and charges for " portion of land revenue due to irrigation ", just what such a peasant would probably have earned in 1926-27. While the case will be supposititious the data on which our calculations are based are those which Government has developed and gathered from the experience of many years' contact with the Punjab peasant. In order that we may be in a position to make certain comparative statements and assumptions due to differences of location of our hypothetical peasant it will aid us to consider the data in three tables relating to the three crops which our peasant is to raise. Our peasant, being interested in producing for the market as most American farmers do, will presumably plant his 9.8 acres in wheat, sugarcane and cotton.

TABLE 29. WHEAT YIELDS PER ACRE IN THE VARIOUS CANAL AREAS OF THE PUNJAB DURING 1926-27 WITH THE GROSS VALUE PER ACRE AND THE WATER RATE ASSESSED ON THE ACREAGE MATURED [1]

Canal	Yield per acre in maunds 1926-27	Price per maund 1926-27	Value per acre of primary produce (wheat)	Value per acre of secondary produce (straw)	Gross value produce per acre 1926-27	Water rate per acre
Western Jumna	12.2	4.7	57	10	67	5-4-0
Sirhind	12.2	4.4	54	9	63	5-4-0
Upper Bari Doab	10.9	4.3	47	8	55	5-4-8
Lower Chenab	12.9	4.6	59	10	69	5-4-0
Lower Jhelum	11.3	4.5	51	9	60	5-4-0
Upper Chenab	10.9	4.5	49	8	57	5-4-0
Lower Bari Doab	11.7	4.5	53	9	62	5-4-0
Upper Sutlej	12.6	4.5	57	10	67	5-4-0
Sidhnai	8.8	4.4	39	7	46	5-4-0
Shahpur	11.3	4.5	51	9	60	5-4-0
Muzaffargarh	6.8	4.1	28	5	33	5-4-0
Upper Jhelum	11.0	4.4	48	8	56	5-4-0

Our average colony peasant is supposed to raise crops which for the year 1926-27 vary according to the Government statistics which we have quoted in the three tables preceding. In the calculations which follow the highest and lowest average production of the three crops will be con-

[1] *Irrigation Report*, 1926-27, *op. cit.*, part ii, pp. 26 *et seq.*

The value of the secondary produce is seen to have varied from Rs. 5 per acre in the Muzaffargarh area to Rs. 10 in the Lower Chenab and two other canal areas. This secondary produce is rarely sold but is used for bullock feed on the peasant's plot. It is probable that in the absence of sufficient straw during an abnormal year the peasant would tend to feed his oxen less rather than to purchase straw. However, to make the return as favorable as possible in the problem of incidence of water rate per acre we shall in each case follow the Government's example and include the value of the secondary produce as indicated. Gross value per acre of wheat according to 1926-27 yields and prices varied from Rs. 33 to Rs. 69 depending upon location of the fields in the different canal areas.

TABLE 30. SUGARCANE YIELDS PER ACRE IN THE VARIOUS CANAL AREAS OF THE PUNJAB DURING 1926-27 WITH GROSS VALUE PER ACRE AND THE WATER RATE ASSESSED PER ACRE MATURED [1]

Canal	Yield per acre in maunds	Price per maund in Rupees	Gross value of yield per acre	Water rate per acre matured
Western Jumna	22.1	Rs. 6.1	Rs. 135	Rs. 12-0-0
Sirhind	24.4	5.7	139	12-0-0
Upper Bari Doab	23.9	5.5	131	12-1-7
Lower Chenab	25.5	5.8	148	12-0-0
Lower Jhelum	15.6	5.0	78	12-0-0
Upper Chenab	19.6	5.4	106	12-0-0
Lower Bari Doab	25.9	5.8	150	12-0-0
Upper Sutlej	25.9	5.8	150	12-0-0
Sidhnai	25.9	5.8	150	12-0-0
Shahpur	15.6	5.0	78	12-0-0
Muzaffargarh	25.9	5.8	150	12-0-0
Upper Jhelum	25.9	8.0	157	12-0-0

TABLE 31. AMERICAN COTTON YIELDS PER ACRE IN THE VARIOUS CANAL AREAS OF THE PUNJAB DURING 1926-27 WITH GROSS VALUE PER ACRE AND THE WATER RATE ASSESSED PER ACRE MATURED [2]

Canal	Yield per acre in maunds	Price per maund in rupees	Value produce per acre 1926-27	Water rate per acre matured
Western Jumna	5.2	Rs. 9.1	Rs. 47	5-8-0
Sirhind	4.4	9.1	40	5-8-0
Lower Bari Doab	2.2	10.5	23	5-8-9
Lower Jhelum	2.4	8.7	21	6-4-0
Lower Chenab	3.7	10.2	38	5-8-0
Upper Chenab	3.2	8.8	28	6-4-0
Lower Bari Doab	3.3	8.7	29	7-8-0
Upper Jhelum	2.6	10.0	26	6-4-0

sidered as possible indicators of the return to the peasant after his year's efforts at agriculture. Since wheat is the

[1] *Irrigation Report*, op. cit., part ii, pp. 26 et seq.
[2] *Irrigation Report*, 1926-27, op. cit., pp. 26 et seq.

chief commercial crop of the Punjab and since wheat plus other small grains account for slightly more than one-half of the cultivated irrigated area in 1926-27 let us suppose that our peasant devoted five acres of his 9.8 acre holding to wheat production. The remaining 4.8 acres in his holding were devoted to the cultivation of American cotton on 2.8 acres and sugarcane on two acres. His gross return from the the three crops and his obligations to the Government are shown in Table 32.

TABLE 32. GROSS RETURN ON AN AVERAGE HOLDING OF 9.8 ACRES OF A PUNJAB PEASANT IN 1926-27 BASED ON MAXIMUM AVERAGE PRODUCTION AND MINIMUM AVERAGE PRODUCTION AS INDICATED IN TABLES 29-30-31. ALSO GOVERMENT REVENUES PAYABLE

Area in acres	Crop	Maximum gross return per acre	Maximum gross return per crop	Minimum average gross return per acre	Minimum average gross return per crop	Water rates per crop according to minimum rates	Land revenue at Rs. 1-8-0 per acre	Portion of land revenue due to irrigation
5	Wheat	Rs. 67	Rs. 335	Rs. 33	Rs. 165	Rs. 26.4		
2.8	Cotton	47	127.6	21	58.8	14.7		
2	Sugar cane	157	314	78	156	24.0		
9.8			776.6		379.8	65.1	14.7	7.0

Gross return from crops to peasant .... Rs. 776.6 or Rs 379.8
Government dues to be deducted [1] 86.8   86.8

Gross return to peasant after payment of Government dues .... 689.8 or 293.0

Table 32 shows that the water rate is not collected on the basis of the amount of produce per acre which the peasant receives but is based on the crop acreage which matures. Thus, for the peasant who happens to be so fortunate as to have a yield which is consistently equal to the highest average for the

IRRIGATION, ITS FINANCIAL ASPECTS 153

province's irrigated areas the incidence of the government charges would not seem to be very heavy, while in the case of the peasant whose yields are consistently equal to the lower average yield for the irrigated areas of the province as a whole, the amount remaining from the gross income after paying the Government dues is still not too seriously impaired. In the former case the Government's share of the gross farm produce is 11.23 per cent while in the latter case it amounts to 22.85 per cent. But the burden of irrigation costs is greater than these figures relating only to gross produce would indicate. It remains to be shown what the peasant's return on his investment in land and stock and implements and for his labor was for the same year.

Before the peasant can tell what his financial position is following a year of labor on his holding he must account for the following costs of production which he will have incurred in bringing his crops to the market: land charges, seeds, wages or customary payments in kind to village menials who have supplied certain of his needs and aided him during the harvest, upkeep of his stock, depreciation in value of stock and implements, interest charges, and finally his own labor. It is evident that the peasant's share of the produce will not be as easily estimated as was the Government's share. Land in the canal-irrigated portions of the Punjab sells at a comparatively high figure as is indicated in the table giving the results of auctions of canal-colony lands in Appendix B. A conservative estimate of the value of the peasant's land would be Rs. 2,940 for the 9.8 acres or Rs. 300 per acre.[1] If he owns his land he should be in a position to earn the prevailing rate of interest on his investment. The Cooperative Credit Societies charge 9 per cent on their loans on

[1] The average price per acre of land sold in the Punjab, accordinf to the 1926-27 *Report of the Department of Agriculture of the Punjab* was Rs. 438 in 1926-27.

agricultural lands. Thus, it is not unreasonable for the peasant to expect to see included in his income for the year an amount which shall include 8 or 9 per cent interest on his investment. This would amount to Rs. 235. Seed sown on his wheat land would have cost him Rs. 22.5 providing one and a quarter bushels of seed per acre at the prevailing price of Rs. 4.5 per maund in most of the canal-colony areas. Sugarcane seed generally amounts to one-twentieth of the outturn per acre, so his two acres of sugarcane would require at least two maunds at a cost of Rs. 5.8 per acre or Rs. 11.6. Cotton seed at the rate of four seers [1] per acre on 2.8 acres would require 11.6 seers and at the prevailing prices would have cost the peasant 2.6 rupees. Thus his total seed costs would be Rs. 46.7 for the year. Labor costs outside of the peasant's family are difficult to state with any definiteness due to the fact that customary payments differ with the locality. The village generally supports at least two specialists, the blacksmith and the carpenter, who are supported by payments in kind. Our peasant would probably have to give to these and other village menials the minimum of one maund of wheat which would cost him Rs. 4.5. This item will vary in different localities, but it is doubtful whether the peasant can avoid paying at least this amount. No peasant can work his land in the Punjab without bullock power to draw his plow. Hence a pair of bullocks must be purchased and supported. To support a pair of bullocks costs from Rs. 56 to Rs. 168 according to one investigator [2] and from Rs. 173 to Rs. 234 according to the Royal Commission on Agriculture in India.[3] A good pair of bullocks in 1926-27 would cost at least Rs. 250 and the depreciation on such stock is

[1] A seer is equivalent to two pounds.
[2] Stewart, H. R., *Some Aspects Of Batai Cultivation In The Lyallpur District* (Lahore, 1926), pp. 6 *et seq.*
[3] *Royal Commission On Agriculture In India, op. cit.*, p. 698.

TABLE 33. AGRICULTURAL IMPLEMENTS REQUIRED BY A FARMER CULTIVATING LAND WITH A PAIR OF PLOUGH CATTLE[1]

Name of implement	Estimated price	Estimated period of usefulness	Used for	Cost per year
Block	Rs. 3-0	3 years	ploughing	Rs. 1- 0-0
Coulter	1-0	3 months	"	1- 0-0
Plough share	0-12	6 months	"	0-12
Beam	5-0	3-4 years	"	1- 6
Kur (part which holds plough share)	1-8	3 months	"	3- 0
Wooden yoke	1-8	1 to 1½ yrs.	"	0-12
Flat levelling beam	5-0	5 years	levelling after ploughing	1- 0
Small wooden yoke	1-4	1½ years		
Rake	1-0	2 years	for making ridges	0- 8-0
Scraper	0-8	1 year	cleaning cattle shed	0- 8-0
5 trowels	2-8	2 years	for hoeing	0- 8-0
2 hoes	5-0	3 years	" "	1-10-0
4 sickles	2-0	2 years	reaping	1- 0-0
4 choppers	5-0	4 years	stripping sugarcane	1- 8-0
2 fodder choppers	2-0	2 years	....	1- 0
Axe	1-8	3 years	general use	0-12
Iron shod stick	0-6	3 years	making holes in the ground	0- 2-0
A 7-forked rake	1-8	3 years	winnowing	0- 8-0
Wooden pitchfork	1-8	2 years	in threshing	0-12-0
				Rs. 17-12-0

estimated at 20 per cent per year. A pair of bullocks is generally considered capable of working about 12 acres of land per year. As the peasant whom we are considering has but 9.8 acres of his own land it is possible that he may hire out his bullocks for a portion of each season so as to earn about 1/6th of their upkeep. The share of the upkeep and depre-

[1] Singh and King, *op. cit.*, p. 154.

The writer has checked the various costs of implements and the period of usefulness of each in various studies in different sections of the Punjab and has found the data fairly representative of actual conditions.

ciation on the bullocks which must be met out of produce from the 9.8 acres will probably not be less than Rs. 150 per year for this average peasant. While the Punjab peasant uses only the most elementary types of implements they nevertheless represent an item of expense which must be deducted from the gross proceeds of his land. The accompanying table gives a list of the common implements and tools which the peasant uses in working his land. Since most of the implements listed in the table are necessary to enable the peasant to carry on his work it appears that Rs. 15 per year would be about the minimum expense which he could expect to incur on this account. Thus the peasant's balance might be indicated as follows at the end of the year 1926-27:

Gross income	Expenditure	
Sale of produce during the year Rs. 776.6 or Rs. 379.8 [1]	Government dues Rs.	86.8
	Seed wheat	22.5
	Seed sugarcane	11.6
	Cotton seed	2.6
	Labor hire	4.5
	Bullock upkeep	100.0
	Depreciation of bullock team	50.0
	Implements & tools	15.0
Rs. 776.6 or Rs. 379.8		Rs. 293.0

Thus the peasant would have a return for his labor and his capital investment of Rs. 483.6 or Rs. 86.8 depending upon whether he had been working land which had produced the maximum average yield during the year or the minimum average yield as per the calculations at the beginning of this section.

At 8 per cent per year the return on the land investment of Rs. 2940 would amount to Rs. 235. In the case of the higher yielding land the deduction of the interest on his land investment would reduce the net income for his labor to Rs. 248.6 which, if divided among the members of his family

[1] *Cf.* Table 32 *supra*, provided he sold the whole of his produce.

would be about Rs. 50 per capita. Evidently the farmer on the lower yielding land would have had but Rs. 86.8 to divide among the members of his family with no return whatever on his investment in land. It is to be admitted that the average yields would probably not work out as consistently as we have indicated. But the two sums which indicate the net return to the peasant in the two cases probably show the limits between which the incomes of the peasants of the Punjab during 1926-27 tended to fall. This would tend to support the tentative conclusion that the return of the peasant is quite low and that his net return on his labor plus his capital investment does not compare favorably with the Government's return of 14.38 per cent on its investment of capital in the irrigation canals of the Punjab. Unfortunately, accurate studies dealing with the net return of the peasant are not yet available. The few studies which students under the direction of the writer have made in various sections of the canal-irrigated portions of the Punjab tend to bear out the conclusion that the peasant's net surplus after paying all costs of production and the Government dues in the form of water rates and land revenue leave but a small income for the peasant's labor and often no return for his capital investment in land, implements and cattle. Apparently, the peasant receives little more than sufficient to permit him to carry on his industry on the barest minimum of subsistence. The average standard of living of the Punjab peasant is thus necessarily low and the income which he receives even when cultivating the irrigated soil is not enough to permit him to raise his standards very greatly.

This point was raised by the Royal Commission on Agriculture in India during its visit to the Punjab in search of evidence as to the status of irrigation and agriculture in the province. The question asked was whether the Government's share of the return from irrigated land in the Punjab was not higher than the standard of living of the peasant, from whom it was collected, justified.[1] The discussions before the Commission brought forth evidence to support at least some of the conclusions of this monograph, namely, that the irrigation canals of the Punjab, as developed by the British Government have been paid for out of current earnings of the canals in operation; that the people of the Punjab have thus not only paid for the development and operation of the canal system but have paid into the Government treasury in various payments for irrigation service a sum almost twice the original capital cost of the canal system plus all operating expense. To be more specific on this point, the capital cost of canal construction in the province amounted to Rs. 32,-16,47,494 while the Irrigation Department has collected from the people of the Punjab a net surplus over and above operating and maintenance expense of the canals during the period of their existence no less than Rs. 58,27,01,089 or Rs. 26,-00,53,595 over and above capital costs, interest charges and operating expenses.[2] That the Government has thus reaped a fair return on its investment is beyond doubt. The suggestion of Mr. White[3] that if the productive irrigation works of the Punjab were to be constructed today they would prove to be unproductive is quite beside the point. Furthermore, his statement is not to be accepted without further investigation. If he wished to go into such an hypothetical discussion he would necessarily have to consider the land which is now irrigated as still lying waste and not under cultivation. It is by no means evident that the projection of irrigation schemes efficiently constructed by means of the modern methods now employed in such works as the Sutlej Valley

---

[1] *Royal Commission On Agriculture, op. cit.*, questions 43,443-43,448.

[2] *Royal Commission Report, op. cit.*, question 43,444.

[3] *Ibid.*, question 43,446. Mr. White, an Irrigation Department official, gave evidence on canal irrigation to the Agricultural Commission.

IRRIGATION, ITS FINANCIAL ASPECTS

Project, would not prove productive works even more quickly than the schemes of the past seventy-five years have been. And, given a population approximately what it is today, the demand for land would certainly cause the bidding for land in the hypothetically-to-be-opened-canal-tracts to bring a good sum to the Government which could be used to aid in financing the costs of irrigation developments. But the case, as a member of the Commission pointed out, is hypothetical and we have been discussing irrigation canals as a going concern of the Government of the Punjab in this monograph.

It is, however, doubtful whether the Punjab Peasant would receive from a reduction in water rates and the other charges accruing from the irrigation services, aid, comparable with that which he now receives from the Government's expenditure of the surplus irrigation revenues which now flow into the general treasury of the Government. In a previous section of this chapter it was shown that the 1927-28 budget estimate of the net revenue from irrigation was Rs. 4,67,-42,000. Further examination of the estimates of the 1927-28 budget expenditure shows the following significant items [1] in which the Punjab peasant is vitally interested:

For capital expenditure on new irrigation works...	Rs. 1,60,71,000
To scientific departments	30,000
To education	1,55,66,000
Medical department	47,36,000
Public health	20,82,000
Agriculture	54,50,000
Industries	8,77,000
Buildings and roads	2,00,26,000
	Rs. 6,48,38,000

A total of Rs. 6,48,38,000 budgeted for services to which the welfare of the Punjab peasant is closely related, a sum incidentally, which more than accounts for the total of the net returns accruing to the Government from irrigation

[1] *Indian Yearbook, op. cit.*, p. 137.

services of the year. While still insisting that Government's assessments on account of irrigation services are high relative to the income which remains to the peasant, amounting to approximately the economic rent on some of the irrigated lands, the expenditure for the benefit of the people of the sums thus collected, appears to justify the high charges for irrigation services. The general standard of living in the Punjab, which is still very low, tends to limit the social outlook of the inhabitants. It is doubtful, therefore, whether the welfare of the people as a whole would be much enhanced by a reduction in irrigation costs, which, while giving the peasant a greater private income, would necessarily take from him the Government supplied services. This conclusion has recently been stated by another investigator whose book has been published during the period in which this monograph was being prepared for the publisher.

It must be definitely recognized that general prosperity in India can never be rapidly or substantially increased so long as any increase in the income of individuals is absorbed not by a rise in the standard of life, but by an increase in the population. The population problem lies at the root of the whole question of India's economic future, and it is useless to bilk (*sic*) the fact.[1]

---

[1] Anstey, V., *The Economic Development Of India* (London, 1929), p. 474. Much of the concluding chapter of this book supports the conclusions which have been formulated in this monograph.

CONCLUSIONS

We have thus reviewed the development of canal irrigation in the Punjab during the British occupation of the province. The first part of this study was devoted to the task of tracing the historical development of the gradually unfolding irrigation schemes of the British officials. The skill with which the network of canals has been successfully constructed, the steady progress achieved in bringing successive areas under the beneficent service of canal irrigation, the efficient manner in which the whole vast series of projects has been developed and the amazing manner in which every obstacle has been surmounted by the British officials and engineers must win the admiration of every student of this development. In 1927-28 an area of 11,157,624 acres had been successfully reclaimed from the desert.

In the second part of the study attention was directed toward certain economic aspects of canal irrigation in the Punjab. It was shown that irrigation projects have been developed at a more rapid rate than the population has increased, resulting in a slight broadening of the agricultural base upon which the people of the province must depend for their existence. There appears to be relatively more food produced per inhabitant in the Punjab today than was possible at the time the British first won possession of this territory. Furthermore, it was shown that temporarily at least, the extension of the cultivated area has tended to relieve the pressure of population on the soil. However, it was also shown that this may prove but a passing benefit, equivalent merely to postponing the necessity on the part of the people of the Punjab to face the population problem and seek its real solution for themselves. A slight lessening of the pressure of population on the more thickly settled portions of the province has occurred following the development of colonization schemes in the newly irrigated sections of the Punjab but it is probable that these slight gains in lessening population pressure in the older sections of the Punjab will have been submerged by natural population increase by 1931. While fragmentation of holdings still occurs in the older portions of the province and is already cutting down the size of the holdings in the canal-irrigated tracts as well, the average cultivator's holding in the canal-colony tracts is still somewhat larger than in the province as a whole. That the canal colony tracts support a relatively more prosperous population than that of the rest of the province was shown by references to the rapidly growing towns in those areas and the increasing density of population in the newly irrigated districts of the Punjab. However, this study has also shown clearly that even in the canal colonies the average yield per acre of the staple crops is lower than one would naturally expect from irrigated land. This is probably due to the traditional methods of cultivation still pursued by the peasants of the Punjab. These methods tend to change very slowly due to the peasant's insistent problem of earning a living from the soil on relatively small holdings. It was shown that the village is still the typical agricultural unit and except in some of the newer canal colonies, the village continues to present an alarming health hazard to its inhabitants. The paucity of modern agricultural implements and the relatively inefficient means at the disposal of the Punjab peasant prevents him from reaping the full benefits from his irrigated land and his labor. Nor can he hope to continue to produce larger yields from his small holdings as long as he pays no attention to the need of fertilizing his fields. However, a brief study of production statistics for some of the staple crops of the Punjab failed to indicate any falling-off in the productiveness of the land during the last twenty-five years. This may, however, have been due to the fact that new tracts have been brought under cultivation in the irrigated sections and the land immediately adjacent, or to the existence of slightly larger fields in the canal-irrigated portions of the Punjab.

The concluding chapter was devoted to a study of the financial aspect of canal irrigation in the province. It was shown quite conclusively that irrigation has been a fiscal success in as much as the entire capital outlay for the irrigation schemes of the British in the Punjab has been entirely repaid out of net earnings and a considerable surplus has been developed in addition. The net surplus above expenses of operation and retirement of debt incurred in the development and operation of canal-irrigation flows into the general revenues of the Punjab Government. It was shown that irrigation receipts now constitute the largest single source of revenue for the Punjab Government. If irrigation receipts were lower than they are it is probable that taxes in general would need to be raised, Government expenditures on beneficent schemes greatly reduced, or perhaps entirely eliminated, or the general administrative efficiency of the Government might be severely reduced. Our study of irrigation costs to the peasant of the Punjab who finally pays for the whole irrigation development and administration through water rates and other incidental charges on account of irrigation services has indicated that the peasant is taxed quite heavily when his low standard of living is considered. The peasant's net return from his land investment and from labor on his small holding is low indeed. It is doubtful, however, whether a lowering of the various Government assessments would aid the peasant to the extent that the Government is aiding him through a serious attempt to increase the social welfare by means of its policy of spending the revenues which it collects from the people of the Punjab for developments of undoubted social value.[1] It is conceivable that the British Government in the Punjab is spending the revenues which in themselves appear heavy for the largest benefit of the people of the Punjab. The vast irrigation projects which the British have developed in the Punjab depend upon political security and efficient administration for their continued operation. This security is dependent upon a strong government. It is thus probable that the cost of irrigation to the people of the Punjab is not unreasonably high if the very profitableness of the irrigation schemes will perpetuate the *Pax Britannica* for the future. The future economic progress of the Punjab is closely related to the efficient administration of the irrigation projects now in operation and projected for future development. The province has much to gain from a continued cooperation with the British in the economic development of its potential resources.

Undoubtedly the greatest monument of English engineering skill in India is not her system of railways, magnificent though they be and remarkable for the economy and safety with which they are worked, but the vast network of canals and channels which distribute water to the fields and enable rich crops to be grown on what were formerly barren wastes.[2]

As suggested in the preface, canal irrigation in the Punjab is a great experiment on the part of a Western Government which is applying the science and initiative at its disposal toward the development of the latent economic resources of an Eastern people whose interest in economic progress perhaps needs still to be greatly enlarged. Intimate contact with western initiative and skill in practical developments of this kind will probably arouse the East from its long lethargy in the economic realm.

[1] 43.5 per cent of the 1927-28 budget was to be applied to further irrigation development, good roads, public health, agriculture and industries and education by the Punjab Government.

[2] Chatterton, Sir A., *Industrial Evolution In India* (Calcutta, 1912), p. 341.

APPENDIX A

RAILROAD MILEAGE AND DATE OF CONSTRUCTION OF IMPORTANT LINES IN THE PUNJAB 1865-1911 [1]

Main Line:—
- Lahore to Multan .... 218 miles opened in 1865
- South to Lahore .... 231 " " " 1870
- Lahore to the West .... 418 " " " 1878
- Lahore to North .... 242 " " " 1880

Branch Lines:—
- Golra-Basal .... 47 " " " 1881
- Amritsar-Pathankot .... 67 " " " 1884
- Rajpura-Bhatinda .... 107 " " " 1889
- Sind-Sagar .... 342 " " " 1890
- Sialkot-Jammu .... 36 " " " 1890
- Raewind-Ferozpur .... 33 " " " 1892
- Southern Punjab [1] .... 400 " " " 1897
- Narwana-Kaithal .... 23 " " " 1899
- Kundian-Campbellpur .... 120 " " " 1899
- Ferozpur-Bhatinda .... 55 " " " 1899
- Wazirabad-Khanewal .... 201 " " " 1900
- Ludhiana-Jakhal .... 79 " " " 1901
- Kalka-Simla .... 59 " " " 1903
- Ludhiana-Macloed ganj .... 152 " " " 1906
- Jech-Doab .... 149 " " " 1906
- Shahdara-Sangla .... 55 " " " 1907
- Khanewal-Lodhran .... 56 " " " 1909
- Amritsar-Patti [3]-Kasur .... 54 " " " 1910
- Kasur-Lodhran .... 208 " " " 1910
- Khanpur-Chachran .... 22 " " " 1911
- Chichoki-Shorkot Road .... 136 " " " 1911

[1] Main Line.

[2] Amritsar-Patti 1906.

[3] Calvert, *op. cit.*, p. 53.

By 1918-19 railway mileage in the Punjab amounted to 5,341, representing a capital outlay of Rs. 10,016 *lakhs* and net earnings for that year of 6.64%. *Ibid.*, p. 55.

APPENDIX B

Table of Results of Auctions of Colony Land [1]

Name of colony	Year of sale	Area sold in acres	Average price per acre
			Rs. A. P.
Chenab (Rakh Branch)	1892	8,783	43- 6-0
Chunian	1896	10,913	50- 5-0
Chenab (Rakh Branch)	1899	5,101	134- 0-0
Chenab (Gugera Branch)	1900	9,913	110- 0-0
Jhelum	1902	4,783	153- 0-0
Chunian	1905	1,132	266- 4-0
Upper Chenab Canal (Gujranwala)	1915	2,185	156- 0-0
Upper Chenab Canal (Lyallpur)	1915	9,393-3-2	179- 0-0
Upper Chenab Canal (Rakh, Lundianwala, Gugera Branch)	1920	6,869-1-2	793- 0-0
Lower Bari Doab Colony	1914	19,271	275-12-0
" "	1915	10,503	130- 0-0
" "	1916	14,442	179- 0-0
" "	1917	8,570	229- 0-0
" "	1918	8,979	294-12-0
" "	1919	8,868	493- 0-0
" "	1920	3,465	593- 0-0
" "	1921	2,591	303- 0-0

[1] *Canal Colony reports*, 1892-1921.

APPENDIX C

Agricultural Experiment Farms of the Punjab Government with Acreage, Located in the Canal Colony Areas [1]

No.	Farms	Area in Acres
1	Sargodha Experimental and Seed Farm	500
2	Chillianwala Seed Farm (District Gujrat)	250
3	Gujrat Government Demonstration Farm	50
4	Lyallpur Agricultural Experimental Farm	671
5	Students' Farm	78
6	Risalewala Cotton Research Farm	200
7	Botanical Section Experimental Research Farm	120
8	Risalewala Seed Farm	878
9	Montgomery Seed Farm	250
10	Raewind (Chunian Colony) District Board Demonstration Farm	46
11	Kahuta (Bara Farm)	558
12	Shergarh (District Montgomery) Seed Farm	275
13	Convillapur Farm Montgomery for growing pure seeds of cotton and wheat (October, 1915)	3,000
14	British Cotton Growing Association, Khanewal for growing pure seed of cotton and asserting marketing of improved vanities (August, 1920)	7,500
15	Sirdar (now Hon'ble Sir) Jogdendra Singh's Farm, Iqbalnagar (District Montgomery). To collect data on how far steam cultivation can be economically employed in the Punjab and to grow and experiment with cotton and such other seeds as the Agricultural Department may designate from time to time	2,000
	Total for Colony Districts	16,376 acres
	Total for the whole Punjab	17,565 acres

[1] Department of Agriculture Report, 1926-27, *passim*.

APPENDIX D

TABLE GIVING DATA RELATING TO THE MILEAGE, DATE OF CONSTRUCTION, COMPLETION, AREA IRRIGATED IN BRITISH TERRITORY AND RIVERS FROM WHICH WATER IS DRAWN, OF CHIEF CANALS OF THE PUNJAB

Canal	Miles of Main canals and branches	Distributaries in miles	Date of opening	Date of completion	Culturable area commanded	Area irrigated in 1926-27	River from which water is taken
Western Jumna	299	1,750	1820	1886	2,304,887	825,669	Jumna
Sirhind	318	1,642	1883	1887	2,093,161	1,659,323	Sutlej
Upper Bari Doab	325	1,548	1859	1879	1,504,059	1,276,495	Ravi
Lower Bari Doab	132	1,233	1913	1917	1,515,524	1,220,034	Ravi & Chenab
Upper Chenab	173	1,268	1912	1917	1,479,890	501,240	Chenab
Lower Chenab	427	2,306	1887	1899	2,629,524	2,562,136	Chenab
Upper Jhelum	128	656	1915	1917	555,652	302,707	Jhelum
Lower Jhelum	181	1,005	1901	1917	1,239,597	881,081	Jhelum
Sidhnai	68	249	1886	1886	395,629	332,489	Ravi
Muzaffargarh	447	665	....	1896	647,317	355,097	Indus & Chenab
Chenab Inundation	221	135	....	1895	386,023	212,404	Chenab
Sutlej Valley [2]	356	506	1926	incomp.	899,023	338,019	Sutlej
Shahpur	116	118	1870	1871	116,286	74,378	Jhelum
Ghaggar	44	34	1897	1896	72,978	14,896	Ghaggar
Indus	418	267	....	1850	649,286	238,645	Indus

[1] *Irrigation Report*, 1926-27, *op. cit.*, part ii, *passim*.

[2] Includes the Upper Sutlej inundation canals.

APPENDIX E

Productive and Unproductive Canal Irrigated Areas in the Punjab from 1887 to 1927 [1]

Year	Area irrigated in acres					
	Kharif [2]	Rabi [3]	Total	Kharif	Rabi	Total
1887–88	645,846	776,929	1,422,775	505,095	413,233	918,328
1888–89	715,632	1,034,325	1,749,957	571,558	425,870	997,428
1889–90	661,677	1,201,779	1,863,456	542,112	452,598	994,710
1890–91	757,535	1,275,965	2,033,500	520,408	462,548	982,956
1891–92	697,454	1,446,938	2,144,392	454,290	596,664	1,049,154
1892–93	823,235	1,024,647	1,847,882	491,151	588,558	1,079,709
1893–94	711,001	1,192,653	1,903,654	409,917	548,747	958,664
1894–95	703,930	1,151,928	1,855,858	500,248	563,048	1,063,296
1895–96	985,827	1,866,556	2,852,383	477,957	461,243	939,200
1896–97	1,534,969	2,114,334	3,649,303	533,745	426,365	960,110
1897–98	1,777,565	2,275,724	4,053,289	565,582	592,778	1,158,360
1898–99	1,413,822	2,458,679	3,872,501	453,402	530,862	984,264
1899–00	1,888,398	2,496,574	4,384,972	520,893	338,629	859,522
1900–01	2,354,174	2,288,678	4,642,852	669,563	688,136	1,357,699
1901–02	1,686,043	2,805,795	4,491,838	508,416	573,105	981,521
1902–03	2,071,363	2,792,911	4,864,274	424,819	390,871	815,690
1903–04	2,450,739	3,160,048	5,610,787	316,418	463,129	779,547
1904–05	2,381,786	3,307,346	5,689,132	296,636	320,616	617,252
1905–06	2,291,460	3,868,911	6,163,371	298,279	452,788	751,067
1906–07	2,435,219	3,467,776	5,902,995	295,898	507,419	803,317
1907–08	2,226,291	3,289,095	5,515,386	258,074	266,484	524,558
1908–09	2,695,518	3,689,522	6,385,050	442,088	526,008	968,096
1909–10	2,285,870	3,831,134	6,117,004	426,020	543,587	969,607
1910–11	2,254,333	4,016,424	6,270,757	459,270	497,015	956,285
1911–12	2,755,324	4,714,321	7,469,645	402,641	459,606	862,247
1912–13	2,929,309	4,531,942	7,461,251	467,957	500,173	968,130
1913–14	3,069,950	4,281,771	7,351,721	464,872	506,871	971,743
1914–15	3,184,751	4,698,877	7,883,628	466,107	613,306	1,079,413
1915–16	2,886,578	5,095,954	7,982,532	362,607	539,672	902,279
1916–17	3,441,467	5,052,400	8,493,867	452,348	633,609	1,085,957
1917–18	3,333,822	4,652,566	7,986,388	463,729	613,784	1,077,513
1918–19	3,465,286	4,882,860	8,348,146	313,334	354,504	667,838
1919–20	4,215,942	5,292,488	9,508,430	455,083	493,145	948,288
1920–21	4,129,503	5,376,130	9,505,633	407,910	360,147	768,057
1021–22	4,277,719	6,124,195	10,401,914	368,128	475,831	843,959
1922–23	4,238,001	6,201,681	10,439,682	429,594	586,199	1,015,793
1923–24	4,137,687	6,226,477	10,364,164	207,525	207,729	415,253
1924–25	4,457,267	5,829,543	10,286,810	209,150	293,486	502,638
1925–26	4,717,519	5,758,481	10,476,000	378,676	254,818	633,494
1926–27	4,716,085	5,765,379	10,481,464	356,701	319,459	676,160

[1] The figures for this table are taken from tables on pages v and iv of the *Irrigation Department, Punjab, Administration Report*, 1926-27.

[2] *Kharif*, the hot weather growing season. Kharif crops are sown in the spring months, April, May, June and harvested in the fall. These crops get the advantage of the scanty natural rainfall.

[3] *Rabi*, the cold weather growing season. Rabi crops, rice wheat or barley are sown in the fall months, September, October, November and harvested in the spring months of March and April. Cotton is a Kharif crop as is also sugarcane and rice.

APPENDIX F

Banks in the Colony Districts

Place	Name of Banks
Bhera	Lyallpur Bank, Punjab & Kashmir Bank.
Chiniot	Lyallpur Bank.
Gojra	Punjab National Bank, Imperial Bank of India.
Gujranwala	Punjab & Kashmir Bank, Imperial Bank of India, Peoples Bank of Northern India, Punjab National Bank, Punjab and Sind Bank.
Gujrat	Lyallpur Bank, Peoples Bank of Northern India.
Hafizabad	Punjab National Bank.
Toranwalla	Punjab National Bank.
Jhang	Lyallpur Bank, Peoples Bank of Northern India, Punjab National Bank.
Khanewal	Peoples Bank of Northern India.
Lyallpur	Allahabad Bank, Central Bank of India, Colony Bank (Head Office), Imperial Bank of India, Lyallpur Bank, Punjab National Bank, Punjab and Sind Bank, Punjab Zamindar Bank (Head Office).
Montgomeey	Imperial Bank of India, Punjab National Bank.
Okara	Punjab National Bank.
Sangla	Punjab National Bank.
Sargodha	Imperial Bank of India, Punjab National Bank, Union Banking & Commercial Association.
Sheikhupura	Peoples Bank of Northern India.

APPENDIX G

NEW AREAS TO BE IRRIGATED BY PROJECTED CANALS IN THE PUNJAB AND BY EXTENSION OF EXISTING PROJECTS [1]

Canals.	Gross area commanded	Culturable area commanded	Area proposed to be irrigated annually
Sutlej Valley Project (new areas)...	5,000,000	4,807,313	3,600,329
Pir Mahal Extension..............	38,443	35,000	24,000
Khikhi Extension	20,525	17,738	12,605
Baralo Branch Extension	173,251	160,000	88,750
Gajargola Distributary Extension ...	11,000	6,025	2,410
Thal Canal	4,599,739	2,596,632	2,170,000
Bhakra Dam Project..............	4,667,047	4,600,000	2,450,000
Jalalpur Hydro-Electric Project	167,824	134,258	100,695
Haveli Project	1,161,164	1,161,164	714,256
Lower Bari Doab Extension	15,000	14,000	11,000
Jassowala Distributary Extension...	60,000	55,000	40,000
Extensions of Other Distributaries of Existing Canals	48,615	48,615	32,000
Total	15,962,608	13,436,745	9,246,055

[1] *Irrigation Report*, 1926-27, *op. cit., passim.*

APPENDIX H

TABLE SHOWING REVENUE FROM MAJOR CANAL WORKS OF THE PUNJAB
(in rupees) [1]

Year	Capital Outlay Direct and Indirect to End of Year	For the Year — Gross Revenue	For the Year — Working Revenue	For the Year — Net Revenue	Percentage on capital
1887-88	5,76,25,823	40,05,923	17,01,248	23,04,675	3.99
1888-89	5,89,48,400	43,64,994	19,72,987	23,92,007	4.06
1889-90	6,00,98,478	49,66,797	21,14,115	28,52,682	4.74
1890-91	6,24,71,861	58,30,440	21,73,686	36,56,754	5.85
1891-92	6,70,91,026	60,40,990	22,51,938	37,89,052	5.65
1892-93	6,97,92,824	68,09,149	24,51,631	43,57,518	6.24
1893-94	7,34,89,876	53,50,907	24,45,060	29,05,847	3.95
1894-95	7,64,95,573	59,65,551	26,50,549	33,15,002	4.83
1895-96	8,02,57,067	70,98,254	28,40,474	42,57,780	5.31
1896-97	8,41,00,265	1,09,44,448	30,87,005	78,57,443	9.34
1897-98	8,70,19,865	1,21,88,419	31,86,227	90,02,192	10.35
1898-99	8,92,16,631	1,24,07,093	35,51,505	88,55,588	9.93
1899-1900	9,09,47,216	1,34,14,661	41,28,714	92,85,944	10.21
1900-01	9,20,75,977	1,46,16,136	43,01,651	1,03,14,485	11.20
1901-02 [1]	8,04,41,638	1,46,35,198	43,26,919	1,03,08,279	11.52
1902-03	10,34,61,088	1,66,14,594	48,22,673	1,17,91,921	11.39
1903-04	10,69,07,947	1,85,73,691	57,18,671	1,28,55,020	12.02
1904-05	10,87,88,346	1,96,10,593	62,38,247	1,33,72,346	12.29
1905-06	11,01,51,719	1,38,36,568	65,90,266	1,17,46,302	10.66
1906-07	11,11,16,550	2,21,75,797	69,78,097	1,51,97,700	13.68
1907-08	11,23,83,171	2,02,79,699	71,79,407	1,31,00,292	11.64
1908-09	11,28,02,325	2,11,00,012	72,71,714	1,38,28,298	12.26
1909-10	11,33,18,773	2,23,99,978	79,19,245	1,44,80,733	12.78
1910-11	11,43,31,721	2,34,16,942	78,02,291	1,56,14,651	13.66
1911-12	11,62,22,814	2,71,10,441	83,56,458	1,87,53,983	16.18
1912-13	14,62,99,761	3,27,93,330	86,81,929	2,41,11,401	16.48
1913-14	16,77,87,618	3,47,39,517	88,11,016	2,59,28,501	15.45
1914-15	17,17,13,716	3,45,62,445	93,24,449	2,52,37,996	14.70
1915-16	21,69,51,444	3,45,89,804	1,09,31,516	2,36,58,288	10.90
1916-17	22,06,16,193	3,96,31,912	1,09,70,471	2,86,61,441	12.99
1917-18	22,33,05,164	3,78,03,265	1,11,03,851	2,66,99,414	11.95
1918-19	22,45,11,551	4,13,14,942	1,10,14,431	3,03,00,511	13.50
1919-20	22,44,57,063	4,47,63,988	1,23,59,237	3,24,04,451	14.43
1920-21	22,52,22,316	4,68,41,809	1,41,65,110	3,26,79,699	14.95
1921-22	22,70,11,881	5,04,72,891	1,69,39,615	3,35,33,276	14.76
1922-23	22,90,41,104	5,31,29,998	1,67,62,986	3,63,67,012	15.87
1923-24	23,21,02,392	5,56,13,332	1,54,00,295	4,02,13,037	15.85
1924-25	23,25,68,890	6,06,29,871	1,63,75,876	4,42,53,995	16.87
1925-26 [2]	23,15,27,367	6,32,32,131	1,56,70,433	4,75,61,698	20.54
1926-27	23,25,75,769	5,83,42,139	1,69,71,196	4,13,70,943	17.79

[1] *Irrigation Report, op. cit.*, part i, p. vi.

[2] Excludes the figures relating to the Swat River Canal, which now belongs to the North-West Frontier Province. The figures up to 1900-1901 include the figures for that canal.

[3] Excludes the figures relating to Indus Inundation Canals transferred to the so-called "unproductive" classification.

SELECTED BIBLIOGRAPHY

Anstey, V., *The economic development of India* (London, 1929).
Bhalla, R. L., *Economic survey of Bairampur* (Lahore, 1922).
Archer, W., *India and the future* (London, 1917).
Baines, J. A., *Famines in India* (British Empire Series, 1899, Vol. I).
Banerjea, P., *Public administration in ancient India* (London, 1916).
Beazley, J. G. and Puckle, F. H., *Punjab Colony manual*, 2 vols. (Lahore, 1926).
Bellais, E. B., *Punjab rivers and works* (London, 1912).
Bert-Chailley, J., *Administrative problems of British India* (London, 1910).
Blair, C., *Indian famines* (Edinburgh, 1874).
Brayne, F. L., *The remaking of village India* (London, 1929).
Buckle, R. B., *Irrigation works in India and Egypt* (London, 1893).
——, *Irrigation works of India* (London, 1905).
Calvert, H., *Wealth and welfare of the Punjab* (Lahore, 1923).
——, *Agricultural holdings in the Punjab* (Lahore, 1925).
——, *Size and distribution of agricultural holdings* (Lahore, 1928).
Chand, Amir Munshi, *A history of the Sialkor district* (translated by C. A. Roe, Lahore, 1874).
Chatterton, A., *Lift irrigation* (Madras, 1912).
——, *Industrial evolution in India* (Calcutta, 1912).
——, *Rural economics in India* (London, 1927).
Connell, A. K., *The economic revolution of India and the public works policy* (London, 1883).
Cotton, Sir Arthur, *Public works in India* (London, 1854).
Coatman, J., *India in 1926-27* (Calcutta, 1928).
Cunningham, J. D., *History of the Sikhs* (London, 1853).
Darling, M. L., *The Punjab peasant in prosperity and debt* (London, 1925).
Dutt, R. C., *Economic history of British India* (London, 1902).
——, *Famines and land assessments in India* (London, 1900).
Gadgil, D. R., *The Industrial revolution in India* (London, 1924).
Grey, R., *Manual of construction and management of district canals* (Lahore, 1885).
Havell, E. B., *A short history of India from the earliest times to the present day* (London, 1924).
Harris, D. G., *Irrigation in India 1923* (London, 1924).
Hunter, Sir William, *A brief history of the Indian people* (Oxford, 1902).
Harding, S. T., *Operation and maintenance of irrigation systems* (London, 1917).

SELECTED BIBLIOGRAPHY

Hamilton, W. S., *Expenses and profits of cultivation in the Punjab* (Lahore, 1916).
Ibbetson, Sir, D. C. J., *Punjab castes* (Lahore, 1916).
Ilbert, Sir Courtenay, *The government of India* (3rd ed., London, 1915).
Joan, F. C., *Root behaviour and crop yield under irrigation* (London, 1924).
Kale, V. G., *Indian industrial and economic problems* (Madras, 1912).
——, *Introduction to Indian economics* (Poona, 1921).
Latifi, A., *The industrial Punjab* (London, 1911).
Latif, S. M., *History of the Punjab* (Calcutta, 1891).
Lucas, E. D., *Economic life of a Punjab village* (Lahore, 1922).
Maclagan, E. D., *The Punjab and its feudatories* (London, 1892).
Macdonald, J. R., *The awakening of India* (London, 1910).
Macgeorge, G. W., *Ways and works in India* (London, 1894).
Morison, Sir Theodore, *Economic transition in India* (London, 1911).
Moreland, W. H., *From Akbar to Aurangzeb* (London, 1923).
Maine, Sir H. S., *Village communities in the East nad West* (London, 1876).
Myles, W. H., *Sixty years of Punjab prices 1861-1920* (Lahore, 1925).
Moreland, W. H., *India at the death of Akbar* (London, 1920).
Narain, Brij, *Eighty years of Punjab prices* (Lahore, 1926).
Naorji, Dadabhai, *Poverty of India* (London, 1888).
——, *Poverty and unBritish rule in India* (London, 1901).
Napier, R., *Abstract of the report of the Bari Doab Canal* (Lahore, 1851).
O'Brien, A., *Manual of Punjab irrigation* (Lahore, 1921).
Petavel, J. W., *Self government and the bread problem* (Calcutta, 1921).
Pillai, P. P., *Indian economic problems* (London, 1924).
Risley, Sir Herbert, *The people of India* (2nd ed., Calcutta, 1915).
Roberts, W. and Faulkner, O. T., *Text book of Punjab agriculture* (Lahore, 1921).
Ross, D., *The land of the five rivers* (London, 1883).
Sarkar, J. N., *Economics of British India* (Calcutta, 1920).
Shah, K. T., *Wealth and taxable capacity of India* (London, 1924).
Smith, V. A., *Oxford history of India* (2nd ed., Oxford, 1923).
Stewart, H. R., *Some aspects of batai cultivation in the Lyallpur district* (Lahore, 1926).
Stewart, H. R. and Katar Singh, *Account of different systems of farming in the canal colonies* (Lahore, 1927).
Strachey, Sir John, *Finance and public works of India* (London, 1882).
Shirras, G. Findlay, *The science of public finance* (London, 1924).
Thompson, W. P., *Punjab irrigation* (Lahore, 1925).
Thorburn, M. M., *Mussalmans and moneylenders in the Punjab* (London, 1886).

Trevaskis, H. K., *The land of the five rivers* (London, 1928).
Widtsoe, J. A., *Principles of irrigation practice* (London, 1914).
Weld, W. E., *India's demand for transportation* (New York, 1920).
Wilson, A. C., *Life and work in a Punjab district* (London, 1895).
Yusaf Ali, Abd Allah, *The making of India* (London, 1925).

GOVERNMENT OF INDIA PUBLICATIONS

Annual review of irrigation in India.
Indian fiscal commission 1921-22 report.
Imperial gazetteers (especially provincial series, Punjab, 1908).
Census reports 1881, 1891, 1901, 1911, 1921.
Indian cotton committee report 1919.
Statistical abstract of British India.
Prices and wages in India, annual report.
Report of Royal commission on agriculture in India (Calcutta, 1928).
General administration reports.
Irrigation commission report (1901-03).

PUNJAB GOVERNMENT PUBLICATIONS

General administration reports 1849-50 to date.
Irrigation department reports, annually since 1882.
Public works department reports since 1864.
Department of agriculture reports, annually since 1882.
Season and crop reports, annually since 1901.
Department of industries reports, annually since 1908.
Land revenue administration reports.
Annual report on operation of Land Alienation Act XIII of 1900.
Punjab government settlement reports 1868 annually.
Punjab census reports, 1868-1881, 1891, 1901, 1911, 1921.
Annual report of cooperative societies.
Punjab cattle census of 1914.
Annually issued district gazetteers.
Reports of the finance department of Punjab Government.

INDEX

Agricultural prices, 36, 112, 150, 151, 152, 154
Agriculture, 17, 22-23, 30, 32, 37, 50, 66, 92, 94-101, 114, 153, 162-63
Akbar, 35
Alexander, 18
Ambala, 82, 86, 89
Amritsar, 25, 26, 28, 52, 69, 82, 85, 86, 89, 102, 126
Anstey, V., 160
Archeological Department, 18
Aryan, 18, 19
Attock, 102

Baden-Powell, B. H., 103
Bahawalpur, 72, 143
Baine, J. A., 24
Balloki, 135
Bari Doab Canal, 25, 27, 33-34, 36, 41, 46, 48, 58, 64, 77, 115, 124, 132, 150, 151
Beas River, 27, 47
Bengal, 112
Bhakra Dam, 73
Bhera, 92
Bikaner, 72, 143
Blair, C., 24
Board of Economic Inquiry, 102
Bolan Pass, 18
Bombay, 112
Burma, 112

Calvert, H. C., 81, 104, 105
Camel maintenance grants, 71
Cannan, E., 38
Capital, 53
Cattle, 24, 154
Census Report, 77, 78, 79, 80, 82, 84, 85, 90, 91, 92
Central workshop, 130, 144
Chatterton, Sir A., 164
Chenab River, 15, 22, 41, 47, 48, 61, 135, 139
Chiniot, 92
Cholera, 56

Chunian Colony, 58, 69, 70
Colonists, 51-52, 56, 60, 67, 69, 82, 85
Colonization, 51-52, 54, 56, 60, 63-66, 82, 85, 87, 101
Consolidation of holdings, 106, 108, 120
Cooperation, 39
Cotton, 22, 61, 108, 110, 113, 115, 117-118, 151-152
Credit, 23, 103, 156-157
Crop yields, 149-155
Crown waste, 51, 60, 62-63, 139
Cultivated area, 74, 77-80, 87, 93, 102, 111, 161, 162
Cultivators' holdings, 101-102, 152

Darling, M. L., 16, 49, 103
Delhi, 19, 37, 38
Dera Ghazee Khan, 39
Demonstration farms, 119, 167

English soldiers, 33

Famine, 23-25, 32, 38, 43-45, 66, 77, 94, 146
Ferozepur, 82, 86, 89, 102
Feroze Shah, 35
Financial Commissioner, 40
Fragmentation of holdings, 72, 101-102, 106-107, 162-163
Frontier Province, 15
Frontiersmen, 39

Golden Temple, 26
Government Agricultural College, 110, 118
Gujranwala, 82, 85, 92, 102, 109, 110
Gujrat, 82, 86, 89, 92, 102, 109, 110, 111, 138
Gurdaspur, 69, 82, 86, 89, 102, 126
Gurgaon, 102

Hafizabad, 92
Harappa, 18

177

INDEX

Himalayas, 18, 21, 28, 29
Hindu law of inheritance, 72, 106
Hindustan, 18, 38
Hissar, 86, 89, 102
Hoshiarpur, 69, 82, 86, 89, 102
Huslee Canal, 22, 26-28, 30, 33, 34, 48

Ibbetson, D. C. J., 25
Implements, 118-119, 153, 155-156
Indian Irrigation Commission, 24, 44-45
Indian Mutiny, 25, 33, 37
Indus River, 15, 16, 17, 22, 34, 36, 40, 47
Invaders, 17-18
Irrigation Department, 36, 113, 122, 148, 158

Jalalpur Jattan, 92
Jhang district, 57, 82, 85, 90, 92, 102, 109, 110, 138
Jhelum district, 82, 86, 89, 102
Jhelum River, 15, 47, 48, 60, 61, 138
Jullundhar, 69, 82, 86, 89, 102
Jumna River, 22, 34, 42, 73

Kabul, 22
Kale, V. G., 37
Kangra, 102
Kapurthala, 86
Karachi, 57, 117
Karnal, 102
Kashmir, 19, 46, 74
Khanewal, 57
Khanki, 61
Khyber, 18
King, C. M., 100, 155
Kusoor, 28

Labor, 42, 43, 154
Lahore, 17, 19, 25, 26, 33, 58, 61, 69, 82-83, 85, 89, 102, 109, 110, 111, 126
Land Alienation Act, 104
Land revenue, 31, 103, 125, 126, 128
Land values, 104-108, 112, 120, 153, 154
Lawrence, Sir Henry, 19
London, 45
Lord Hardinge, 27
Lord Minto, 19
Lower Bari Doab Canal, 60, 63, 68-69, 71-72, 88, 90, 115, 124, 133, 135, 150

Lower Chenab Canal, 54, 56, 61, 68, 71, 83, 87, 90, 115, 124, 136, 138, 150
Lower Jhelum Canal, 58, 60, 68-69, 71, 87-88, 115, 124, 138, 150
Lower Sohag Canal, 50-51, 64
Loveday, A., 24
Ludhiana, 82, 86, 89, 102
Lyallpur, 57, 61, 82, 90, 92, 102, 109, 110, 111

Madhopur, 26, 144
Madras, 112
Maine, Sir H. S., 20
Malikana, 69
Mangla, 61
Maratha, 19
Merala, 61, 135
Mianwala, 86, 89, 90, 102
Military operations, 37
Mohammedan, 18, 19, 25, 88
Money lenders, 23, 53, 104
Montgomery, 51, 61, 72, 82, 86, 89, 90, 92, 102, 109, 110, 126
Multan, 17, 19, 28, 51, 61, 82, 84, 86, 90, 102, 109, 110
Mussoo Khan, 39, 40
Muzaffargarh, 102
Muzaffargarh canals, 115, 124, 140, 150, 151

Native soldiers, 43
Native states, 45
Navigation, 28, 29, 33, 145, 148
Nazarana, 69, 70
Net revenue, 30, 31, 34, 128, 131-133, 139, 159, 172
Nomads, 57

Occupancy rights, 67

Pakpattan, 92
Pakpattan Canal, 73, 115, 142
Para Canal, 50-51
Peasant, 22-23, 27, 38, 62, 66, 93, 96-98, 103, 106, 108, 119, 120, 128, 148-160, 163-164
Persian wheels, 53
Pigou, A. C., 95
Population, 23, 25, 43, 47-48, 52, 58, 62, 68-69, 77-95, 102, 105, 120-121, 160-163
Primogeniture, 72
Prisoners, 42, 43

INDEX

Private property, 56, 103
Productive works, 44, 145
Proprietary rights, 59, 62, 69
Protective works, 43, 44, 51
Puckle, F. H., 49
Punjabi, 17, 22, 93
Punjab village, 64-67, 91, 93, 96-100, 106, 120, 154, 162
Putiala, 39, 42, 45-46, 86

Railroads, 37, 46, 91, 125
Rajputana, 16
Rainfall, 17, 23
Ramnagar project, 55
Rasul, 60
Ranjit Singh, 19
Ravi River, 15, 26-28, 41, 47, 48, 51, 61, 139
Rawalpindi, 82, 86, 102
Rohtak, 102
Round Table, 46
Royal Commission on Agriculture, 36, 98, 107, 112, 154, 157-159
Roy, S., 24

Service grants, 60, 68, 70-71
Shah Jehan, 26
Shahpur Canal, 115, 146
Shahpur district, 82, 86, 90, 102, 109, 110, 111, 138
Shahpur inundation canals, 58, 74, 150
Sheikhupura district, 57, 61, 82, 102
Sialkot, 61, 82, 85, 86, 89
Sidhnai Canal, 51, 54, 56, 64, 84, 115, 124, 139, 140, 150
Sind desert, 15, 16
Sindhia, 20
Sikh, 19, 26, 27, 43
Singh, G, 100, 155
Sirhind Canal, 41, 43, 45-46, 78, 115, 124, 132, 150
Smith, A., 38
Smith, V. A., 19
Soil deterioration, 112

Standard of living, 48-49, 65, 102, 114, 119, 154-157, 160, 163
Stewart, H. R., 154
Sugarcane, 108, 111, 113, 115, 151-152, 154
Sutlej River, 15, 17, 19, 28, 34, 41, 48, 50, 72-73
Sutlej Valley Project, 72, 123-124, 128, 142, 145, 158

Taxila, 18
Tenants at will, 67
Thompson, W. P., 35
Transportation, 23, 25, 30, 37, 39, 53, 57, 112
Trevaskis, H. K., 17, 20
Triple Canal Project, 60-64, 135-136, 137

United Provinces, 15, 112
Upper Chenab Canal, 60, 63, 69, 88, 115, 124, 134, 136
Upper Jhelum Canal, 61, 63, 70, 87, 115, 124, 137, 150
Upper Sutlej Inundation Canals, 51, 115, 150-151

Vakil, C. N., 44, 127

Water-logging, 35-36
Water power rights, 30, 127, 145
Water rates, 26, 30, 70, 112, 125-127, 145, 148-151
Wazirabad, 57
Western Jumna Canals, 34, 47, 77, 115, 124, 130-131, 150
Western Jumna Canals, 34, 48, 77, 115, 124, 130-131, 150
Wheat, 22, 61, 108-109, 113, 115-117, 150-152
Woolar Lake, 46, 73

Yoeman grantees, 69-70
Young, P., 70

Zamindars, 36, 39, 40

Bei Fragen zur Produktsicherheit wenden Sie sich bitte an:
If you have any questions regarding product safety,
please contact:

Walter de Gruyter GmbH
Genthiner Straße 13
10785 Berlin
productsafety@degruyterbrill.com